D0856852

THE NEW ARGONAUTS

THE NEW ARGONAUTS

Regional Advantage in a Global Economy

AnnaLee Saxenian

HARVARD UNIVERSITY PRESS
CAMBRIDGE, MASSACHUSETTS
LONDON, ENGLAND ❖ 2006

Printed in the United States of America

Library of Congress Cataloging-in-Publication Data

Saxenian, AnnaLee.
The new argonauts : regional advantage in a global economy / AnnaLee Saxenian.
p. cm.
Includes bibliographical references and index.
ISBN 0-674-02201-7 (alk. paper)
1. High technology industries—Developing countries.
2. Immigrants—California—Santa Clara Valley (Santa Clara County)
3. Cooperative industrial research—California—Santa Clara Valley
(Santa Clara County) I. Title.

HC59.72.H53S29 2006
338.4'7004091724—dc22 2005055112

To my parents
Lucy and Hrand Saxenian
with love

CONTENTS

INTRODUCTION

Intel Corporation learned in 1974 that one of its leading scientists, Dov Frohman, planned to return to his native Israel to take a research and teaching position at the Hebrew University of Jerusalem. Frohman was among the company's most talented engineers—he had invented and developed the erasable programmable memory chip (EPROM). To avoid losing him, Intel established its first integrated circuit design center outside of the United States, in Haifa, and asked Frohman to lead it. Thirty years later, Israel had become a leading center of world-wide IC design and manufacturing, renowned for its innovative security and communications products.

In 1984 six Chinese engineers quit their jobs at Fairchild Semiconductor in Silicon Valley to return home and start Taiwan's first three private semiconductor firms. At the time, the only semiconductor company in Taiwan was a government-sponsored start-up. Nasa Tsai, a Stanford Ph.D., recalls that "even though we'd all been in the U.S. for a dozen years, we still had friends, colleagues, and classmates here in Taiwan," which made starting an overseas business feasible. These early semiconductor companies had an uphill fight. One firm, Quasel, folded after five years. The other two, Mosel and Vitelic, survived only by licensing their technology. Neither was able to build its own foundry until the two companies merged a decade later. Mosel co-founder Peter Chen reports that the company nonetheless had an enormous impact: "We upgraded Taiwan's technology by two generations as soon as we returned." Mosel designed Taiwan's first 16K SRAM; Vitelic co-developed a 64K DRAM a year later. The merged Mosel-Vitelic began volume production in the early 1990s and, riding the semiconductor industry boom, became profitable and went public in Taiwan in 1995. These pioneers became role models for hundreds of subsequent returnees, who by the late 1990s helped to make Taiwan the world's third largest producer of integrated circuits.

Ning-San Chang moved to Shanghai in 2002 with two Chinese colleagues from Silicon Valley to start BCD Semiconductor Manufacturing, China's first bipolar IC foundry. Chang had earned a Ph.D. in electrical engineering from Purdue University, and worked for twenty years in the U.S. semiconductor industry. He regretted missing earlier chances to return to Taiwan and felt lucky that he could contribute to developing the mainland Chinese semiconductor industry. In his words: "It won't make much of a difference if I work in Silicon Valley or Taiwan now, but there are so many opportunities in Shanghai now that I know I can make a difference here." BCD, which designs and manufactures low-cost analog semiconductors for use in phones, computers, and consumer electronic products, has received two rounds of venture funding from top-tier investors in the United States and Asia, including Intel Capital and Acer Technology Ventures, and has recruited "best of breed" managers from Silicon Valley and Taiwan, as well as China, to work in its Shanghai operations.

In 1999 National Semiconductor rejected a plan to design a chip for MP3 players. In classic Silicon Valley fashion, the executives and engineers championing the plan quit and raised venture funds to start Portal Player Inc. Three of the co-founders were of Indian

> descent, and they quickly established a design and development team in Hyderabad, India, making the start-up a global company from the start. They hired J. A. Chowdary, one of the chief architects of the Indian information technology industry, to run the operation. Today half of the firm's software engineers are in India, where they contribute to the millions of lines of embedded code—a miniature operating system, or system-on-a chip (SoC)—that control the video, graphics, and music in Apple's popular iPod media player. Portal Player specializes only in intellectual property (IP) and collaborates closely with partners that coordinate the complex manufacturing logistics and quality assurance required to translate its designs into cost-effective integrated circuits. The iPod may be a symbol of American technology and culture, but it is created by a far-flung network of highly specialized enterprises located in Silicon Valley, India, Taiwan, mainland China, and elsewhere.

During the period following World War II, Taiwan and Israel were peripheral economies, yet in the last decades of the twentieth century they emerged as important global centers of innovation and growth. By 1990, output from Taiwanese and Israeli information technology sectors exceeded that of larger and wealthier nations like Germany and France. At the same time, the rapid growth of technology businesses in select urban centers of China and India made these countries the envy of the developing world. The pioneers of these profound transformations in the global economy are the foreign-born, technically skilled entrepreneurs who travel back and forth between Silicon Valley and their home countries. Like the Greeks who sailed with Jason in search of the Golden Fleece, the new Argonauts undertake the risky but economically rewarding project of starting companies far from established centers of skill and technology.

Traditional theories of economic development assume that new products and technologies emerge in industrialized nations that can combine sophisticated skills and research capabilities with large, high-income markets, and that mass production is shifted to less costly locations once the product is standardized and the manufacturing process has matured. In this view, success in the periphery builds on the success of more advanced economies: late developers are destined to remain followers because leading-edge skills and technology reside in the corporate research labs and universities in the core.

This model leaves little room for the independent development of technological capabilities in the periphery. At best, foreign investment from the core might accelerate the mastery of mass manufacturing and incremental upgrading of local suppliers. The primary route to development, in this standard view, is the mobilization by the state, in conjunction with national banks and large-scale industry, of the resources to either replicate or import the manufacturing techniques developed in the core. Indeed, multinational corporations have historically relied on peripheral economies like India and China as sources of low-cost manufacturing skill.

Recent developments in the world economy have undermined the power of this core/periphery model, however. The increasing mobility of highly skilled workers and information, as well as the fragmentation of production in information and communication technology sectors, have led to unprecedented opportunities for formerly peripheral economies. Regions that missed the postwar economic boom have provided fertile environments for decentralized growth based on entrepreneurship and experimentation. The key actors in this process are neither policymakers nor multinational corporations acting in isolation, although both certainly play a role, but rather communities of technically skilled immigrants with work experience and connections to Silicon Valley and related American technology centers.

Communities of U.S.-educated immigrant engineers now routinely transfer up-to-date information and know-how to help their home economies participate in the IT revolution. Capitalizing on their experience and the support of professional networks, these new

Argonauts can quickly identify promising new market opportunities, raise capital, build management teams, and establish partnerships with specialist producers located far away. The ease of communication and information exchange within ethnic professional networks accelerates learning about new sources of skill, technology, and capital as well as about potential collaborators, facilitating the timely responses that are essential in a highly competitive environment.

This is not a one-way process. As recently as the 1970s only large, established corporations had the resources and capabilities to grow international businesses, and they did so primarily by establishing marketing offices or manufacturing branch plants overseas. Today the fragmentation of production and the falling costs of transportation and communication allow even very small firms to build long-distance partnerships in order to tap overseas expertise, cost savings, and markets. Start-ups in Silicon Valley today are often global businesses from their first day of operations; many are able to raise capital only if they have a demonstrated ability to subcontract manufacturing or software development and to market their products or services outside the United States.

In this environment, the scarce competitive resource is the ability to locate foreign partners quickly and to manage complex business relationships and teamwork across cultural and linguistic barriers. This is particularly challenging in high-tech industries in which products, markets, and technologies are continually redefined—and where product cycles are often nine months or less. First-generation immigrants like the Chinese and Indian engineers in Silicon Valley who have the necessary language, cultural, and technical skills to function well in the United States as well as in their home markets have a commanding professional advantage. Ethnic professional associations and networks extend these advantages by enabling new as well as established ventures to quickly identify and build partnerships with distant suppliers and customers.

The economies of Taiwan and Israel developed initially as sources of low-cost skill for labor-intensive manufacturing, just as the old model predicts. However, as U.S.-educated immigrant engineers returned home, either temporarily or permanently, they transferred the institutions of entrepreneurship from American technology regions like Silicon Valley to their home countries. In both countries, a new generation of start-ups pursued the often incremental but cumulatively significant improvements in processes and products that now provide a distinctive competitive advantage. Taiwan is now home to the world's leading IT and networking hardware manufacturers, and Israeli companies have pioneered innovations in network security, telecommunications software, and electronic components. This process is still at an earlier stage in China and India. Yet producers in both countries are quickly leading regions in these nations along upgrading trajectories similar to those in Israel and Taiwan.

The new Argonauts are undermining the old pattern of one-way flows of technology and capital from the core to the periphery, creating far more complex and decentralized two-way flows of skill, capital, and technology. They have created dynamic collaborators in distant and differently specialized regional economies, while largely avoiding head-on competition with industry leaders. Silicon Valley is now at the core of this rapidly diversifying network because it is the largest and most sophisticated market as well as leading source of new technology.

Will Silicon Valley remain the center from which new Argonauts continue to set forth? I will return to this question in the concluding chapter. It is clear that the relationships between these newly emerging technology regions are multiplying, and that the new markets opening up in China and India will further transform the dynamics of the world economy.

Not all peripheral regions can become centers of technology entrepreneurship like Taiwan and Israel. Nations that have invested heavily in higher education—particularly technical education—are the best positioned to compete in these industries. Most of the developing economies in Asia, Africa, and Latin America have failed to make such investments. Others, like Singapore and Scotland, do not lack skill but have focused their development efforts on attracting foreign investment, or, like Iran and Russia, they lack the political stability required to entice emigrants to return as entrepreneurs.

Many advanced industrial nations, such as France and Japan, have been slow to develop institutions that support technology entrepreneurship. The state planners, bankers, and large corporations that collaborated to support the postwar mass production system have resisted such change. Moreover, the economic opportunities at home for talented young people in these countries are sufficiently attractive that few choose to leave for higher education, and if they do, they return home as soon as they graduate to take top corporate or civil service positions rather than staying abroad. This has isolated them from distant sources of technology and markets.

Developing countries that invested heavily in education in the postwar period did not achieve their planned outcome either. It was these countries that suffered from a "brain drain" as their most talented students left to take advantage of educational opportunities abroad. Policymakers complained bitterly about these losses and even sought to control them. At the time, none foresaw that emigrant engineers and entrepreneurs might become a valuable asset in the twenty-first-century global economy.

A small but meaningful proportion of individuals who left their home countries for greater opportunities abroad have now reversed course, transforming a brain drain into a "brain circulation." They are returning home to establish business relationships or to start new companies, while maintaining their professional and social ties to the United States. The spread of venture capital provides an example of this process. In the early 1980s, emigrants returning from Silicon

Valley began to transfer the model of early-stage high-risk investing to Taiwan and Israel. These native-born investors brought cultural and linguistic know-how as well as the capital needed to operate profitably in these markets. They also brought technical and operating experience, knowledge of new business models, and networks of contacts in the United States. Today Israel and Taiwan boast the largest venture capital industries outside of North America, and both support high rates of new firm formation.

When foreign-educated venture capitalists invest in their home countries, they transfer first-hand knowledge of these financial institutions to the periphery. These individuals, often among the earliest returnees, frequently serve as advisers to domestic policymakers as well. Moreover, as engineers and managers return home, they bring the worldviews and identities that grow out of their shared professional and educational experiences. They form technical communities with the potential to jump-start local entrepreneurship, and the most successful returnees build alliances with technical professionals, businesses, and policymakers in their home countries. Not surprisingly, this has proved to be significantly easier to achieve in Israel and Taiwan than in the complex political economies of China and India. Over time, however, the impact of returning entrepreneurs and their communities may be more far-reaching in these two global giants.

In the challenging process of starting companies far from established centers of skill and technology, today's Argonauts depend upon air travel and electronic connections that allow real-time communications around the globe. They are also exploiting the entrepreneurial opportunities generated by the changing organization of IT production, changes that were triggered by the destabilizing emergence of the Silicon Valley system. Most of them are men. According to Apollodorus, the original Argonauts consisted of forty-three men and one woman—a ratio that the new Argonauts are slowly improving.[1]

The experience of the new Argonauts suggests that policymakers in developing economies should devote their efforts to facilitating a bottom-up process of entrepreneurship and long-distance information exchange rather than making top-down attempts to grow the next Silicon Valley. It is better to connect to the Silicon Valley economy by integrating into its technical community than to try to replicate it; better to develop capabilities and products that are complementary to those in Silicon Valley than to compete head-on with well-established companies.

I begin the book by presenting a framework for understanding the new Argonauts and the creation of a Silicon Valley–style regional advantage in once-peripheral economies. The second chapter documents the way foreign-born engineers in Silicon Valley mobilized ethnic social and professional networks that allowed them to succeed as entrepreneurs during the 1980s and 1990s. In Chapter 3, using the case of Israel as an illustration, I describe how these immigrant communities seeded new centers of entrepreneurship in their home countries by transferring the institutions and practices of modern technology regions while also maintaining close ties to Silicon Valley.

The next four chapters examine the impact of Silicon Valley's Chinese and Indian communities on their home countries. Returning entrepreneurs and their communities have not simply tapped low-cost skill; in all three cases (Taiwan, China, and India) they have also transformed the local environment for entrepreneurship by addressing immediate obstacles to success, ranging from capital market and telecommunications regulations to the educational systems and research institutions. In each case, they have contributed to the rapid creation and improvement of local capabilities.

In Chapters 4 and 5 I analyze the evolution of the Silicon Valley–Taiwan connection by tracing the influence of U.S.-educated engineers on technology policy, the establishment of a venture capital industry, and the entrepreneur-led growth and technological upgrading of the domestic semiconductor and electronic systems manufacturing industries in the 1980s and 1990s. Producers in Taiwan's Hsinchu region have flourished not in isolation, but through their ongoing collaborations with distant partners.

In Chapter 6 I turn to Silicon Valley's returning mainland Chinese engineers, arguing that Taiwan has provided a model as well as a significant source of technology and talent for China. The longer-term impact of returning entrepreneurs will depend as much on domestic political and economic reforms as on these cross-regional activities. It is easy to overestimate China's technology capabilities, but they are already the world's IT manufacturing center, and collaboration with leading-edge specialists in Silicon Valley and Taiwan offers great promise. In Chapter 7 I trace the emergence of the cross-regional community of U.S.-educated Indian technology entrepreneurs and managers since the 1980s, describing their contributions to the sustained upgrading of software services and development capabilities in India over the past decade, as well as the enduring constraints of local infrastructure and politics.

In the concluding chapter I discuss the possibilities for expansion of this model to other countries, as firms continue to open their boundaries and cross-regional entrepreneurs help localities deepen their own capabilities. The rise of a network of regional economies with distinct and complementary specializations has the potential to change the nature of global competition, creating opportunities for sustained growth through reciprocal upgrading.

The new Argonauts challenge not only our models of economic development but also the dominant policy agendas and corporate strategies of the last century. By creating dynamic new centers of technology entrepreneurship, they are extending the Silicon Valley system to distant locations faster and more flexibly than either multinational corporations or government bureaucracies could. Theirs is a remarkable journey that demands the attention of scholars, policymakers, business leaders, and citizens seeking to understand how regional advantage shapes global competition.

Finally, this book is not primarily about public policy, although the explosive growth of the Chinese and Indian economies is currently a topic of significant public interest. The story of the new Argonauts is an account of profound transformations in the global economy. For Americans accustomed to unchallenged economic domination, the fast-growing capabilities of China and India can seem threatening. Some conclude that America has been foolish to educate foreigners who steal American technology and export American jobs. While the imagery is vivid, the economic evidence weighs against this view. The Argonauts have made America richer, not poorer. Far from stealing jobs, immigrant entrepreneurs have created them in very large numbers, both in the United States and overseas. America benefits from the dramatic reductions in the cost of both producer and consumer technologies made available by the growth of overseas technology regions, as well as from access to fast-growing foreign markets.

America remains the world's business and technology leader precisely because it attracts talent and capital from around the globe and maintains world-class technology education and research while encouraging intense competition, collaboration, transparency, and entrepreneurship. No other country could have spawned the new Argonauts; none has benefited more from their labors; and none would be hurt more by policy that undermined the openness of the entrepreneurial ecosystem in America's technology regions.

1 ❖ Surprising Success

The recent economic successes in Israel, Taiwan, China, and India are particularly surprising given that all of these countries had poor agricultural economies throughout the postwar boom. China and India are still among the world's poorest countries, and until recently they were isolated from the competitive pressures of the world economy. As recently as the late 1980s, few would have predicted that companies in these backward economies would compete in the leading technology sectors of the global economy. Yet today Taiwan's specialized semiconductor and computer-related firms define the international state of the art for efficient, flexible electronic systems manufacturing and logistics. Israel, with a population of just over 6 million, can claim more than a hundred companies listed on NASDAQ—more than in any other country outside North America. India is the leading global provider of software development services and business process outsourcing, and China has overtaken Japan to become the world's second largest IT manufacturing center outside the United States.

What can explain these surprising successes? Standard accounts of industrialization focus either on the activities of the nation-state or on investment patterns of multinational corporations. In the former, the archetypal development program is an efficient and relatively autonomous national bureaucracy that collaborates with dominant corporations and financial institutions to channel domestic investment and resources to emerging sectors. Export-led industrialization in Japan and Korea involved the early identification and imitation of technologies such as semiconductors and disk drives, followed by the rapid ramp-up and refinement of high-volume, low-cost mass production. Policymakers coordinated the enormous investments required for research and development and for large-scale manufacturing.

The other dominant accounts of economic development give a central role to foreign direct investment, typically motivated by competitive pressures to cut costs. Multinational corporations from the United States, Japan, and Europe aggressively expanded manufacturing in low-wage locations of the Third World starting in the 1960s. These investments ensured the creation of the supplier infrastructures and skill base needed to master high-volume manufacturing of electronic components in places such as Singapore, Malaysia, Scotland, and Ireland. They also contributed to substantial improvements in the standard of living in these economies.

Neither approach explains the rise of indigenous entrepreneurship in the new technology regions. Policymakers have invested, to varying degrees, in infrastructure, education, and research and development, and multinationals have expanded their investments as well. But the innovative dynamism of the new regions has come not from the state, established firms, or foreign investors, but rather from the domestic enterprises started since 1980s with little (if any) state support. The most successful of these, like Taiwan's Acer, Israel's Mirabilis, India's Infosys, and China's Lenovo, have grown to become significant global competitors, but they are only the most visible of the technology businesses in these economies. Each is surrounded by hundreds of enterprises of differing sizes, ages, ownership, and specializations, and by an ongoing flow of entrepreneurial start-ups and failures.

The new Argonauts are the main actors in this process. They are products of the postwar brain drain, when thousands of foreign-born students earned science and engineering degrees at U.S. universities annually, and remained after graduation to work in the nation's fast-growing technology companies. These immigrant technologists—often the best and brightest from their home countries—integrated themselves into local economies by creating ethnic social structures and institutions that supported professional advancement and entrepreneurial success. By extending these social networks to their home countries, they have transplanted the institutions and relationships of technology entrepreneurship and are reshaping global technology competition.

PROBLEMS AND SOLUTIONS

Late-developing economies typically face two major disadvantages: they are remote from the sources of leading-edge technology, and they are distant from developed markets and the interactions with users that are crucial for innovation.[1] Firms in peripheral locations traditionally use a variety of mechanisms to overcome these disadvantages, from joint ventures and technology licensing to foreign investment and overseas acquisitions. However, the existence of a network of technologists with strong ties to global markets and the linguistic and cultural skills to work in their home country is proving to be a more efficient and compelling way to overcome the limitations of distance. Cross-regional entrepreneurs and their communities facilitate the diffusion of technical and institutional know-how, provide access to potential customers and partners, and help to overcome the lack of reputation or information that results in trade barriers for isolated economies.

The increasing sophistication of information and communication technologies and the liberalization of global markets have accelerated this process. It is now quick, simple, and inexpensive to communicate internationally and to transfer information between distant locations. Information systems that facilitate the formalization of knowledge are dramatically expanding the volume as well as the variety of information exchange. However, information technology alone cannot ensure successful coordination or efficient transfers of technical and institutional knowledge. Long-distance collaborations still depend heavily upon a shared social context and language to ensure mutual intelligibility between partners in markets where speed and responsiveness determine competitive success.

Market liberalization has likewise been important to the economic transformation of both China and India. However, the reduction of trade barriers and bureaucratic intervention alone does not create the institutional capability and social context, or the domain knowledge, required for entrepreneurial success in global industries. Technology entrepreneurship remains highly localized even in the most advanced economies, and it cannot be created by fiat, as evidenced by decades of failed attempts to "grow the next Silicon Valley." Efforts to jump-start entrepreneurship by mobilizing researchers, capital, and a modern infrastructure cannot replicate the shared perspective, language, experience, and trust that permit open information exchange, collaboration, and learning (often by failure) alongside intense competition in places like Silicon Valley.

The new technology centers differ significantly from one another, and from Silicon Valley, in their technological sophistication as well as in the specialization of local producers. Cross-regional entrepreneurs rarely compete head-on with established U.S. producers; instead they build on the skills and the technical and economic resources of their home countries. Israeli entrepreneurs, for example, have successfully applied the findings of the nation's advanced military research to innovations in the Internet security and telecommunications arenas. Indian entrepreneurs, by contrast, recognized the opportunity to mobilize thousands of underemployed English-speaking Indian engineers to provide software development services for American corporations. Returning entrepreneurs are ideally positioned to identify appropriate market niches, mobilize domestic skill and knowledge, connect to international markets, and work with domestic policymakers to identify strategies to overcome obstacles to further growth.

Investors and engineers in these regions are promoting the development of local ecosystems for entrepreneurship, while also maintaining close connections to technology and markets in the United States. The infrastructure for entrepreneurship is best developed in Israel and Taiwan, where thousands of technologists have returned since the 1980s. Both regions have also completed several entrepreneurial cycles in which successful entrepreneurs have reinvested their capital and contributed accumulated know-how and contacts to a subsequent generation of technology ventures, while also serving as role models. This cycle is both the cause and the consequence of the relationships and informal information flows that support regional experimentation and learning. It does not guarantee the success of any individual firm, but it provides local producers with the collective capacity to adapt and improve.

The dynamism of these technology regions is not reducible to cost advantages. Investors in India and China may have been initially motivated by the availability of low-cost skill, but the concentration of technology production has already generated rapidly rising wages and intensifying congestion in these regional economies. Engineering salaries in both Bangalore and Shanghai, for example, are now among the highest in their respective nations, yet new and established producers continue to cluster there rather than seeking lower-cost locations. The experience of Silicon Valley demonstrates that decentralized economies can flourish long after their labor cost advantages disappear, as long as local investors and entrepreneurs are able to create and recreate their regional advantage by collectively learning, innovating, and upgrading local capabilities.

The contributions of an international technical community in transferring the institutions of technology entrepreneurship should not be confused with the broader role of a diaspora in the home country. The aggregate remittances, investments, or demonstration effects of a diaspora can affect an economy in a variety of different but largely limited ways. The new Argonauts, by contrast, are a small subset of highly educated professionals whose potential contributions to economic development are disproportionately significant. They are not typically drawn from the traditional economic or political elites of their home countries. Instead, they are often the top engineering students from middle-class households whose access to education in the United States has exposed them to a very different technological and institutional environment.

At one level, this is not surprising. Long-distance migrations have shaped the contours of the world economy throughout history. The transfer of skill and know-how that accompanies the movement of individuals and groups within and between nations can have an enduring impact on patterns of economic development, as with the modernization of Japan during the Meiji restoration in the nineteenth century, or the transfer of British textile and German steel technology to the United States during the nineteenth century. Economic historians have documented the contributions of personnel recruitment to knowledge transfer and have demonstrated that the experience, relationships, and tacit knowledge that reside in individuals and their communities play a central role in long-distance transfers of technology and economic institutions.[2] But tacit knowledge alone is not sufficient. Transferring production to a new location requires deep knowledge of the local context—the subtle as well as the more apparent differences in social, cultural, and institutional settings. And long-distance collaboration rarely succeeds without the shared language and social context that facilitate communication. Because there are few substitutes for native experience, cross-regional entrepreneurs accelerate the adaptation of technology and institutions to local circumstances that are inevitably different from those in the United States.

FROM BRAIN DRAIN TO BRAIN CIRCULATION

Scholars have long viewed the postwar brain drain as reinforcing global inequality—a net loss for developing economies but yet another advantage for already rich and industrialized nations. According to a classic textbook on economic development, "The people who migrate legally from poorer to richer lands are the very ones that Third World countries can least afford to lose—the highly educated and skilled. Since the great majority of these migrants move on a permanent basis, this perverse brain drain not only represents a loss of valuable human resources but could also prove to be a serious constraint on the future economic progress of Third World nations."[3]

Although data on the emigration of skilled workers are hard to find, the United Nations estimated a total of 300,000 highly skilled emigrants moving from all developing countries to the West during the 1960s, and the 1990 U.S. Census showed 2.5 million highly skilled immigrants, excluding students. Researchers seeking to quantify positive feedback effects from migration, such as remittances, investments, or return migration, concur that these effects have been limited, particularly among highly skilled immigrants who rarely returned to their home countries.[4] Such analyses have even led economists to consider a system of global taxes on either the migrants or the receiving countries to compensate for the loss of human capital by poor nations.[5]

The picture today is different. The countries that suffered most from the postwar brain drain—especially Taiwan, Israel, India, and China—are those that now benefit most directly from the rise of the new Argonauts. Developing nations that invested heavily in high-quality tertiary technical education were most likely to lose their most promising young people to higher education abroad during the postwar years; they also typically lacked the industrial base to employ the larger numbers of graduates who never left the country. As a result, these countries were often home to large numbers of unemployed or underemployed graduates. In recent decades U.S.-educated engineers from places like Taiwan and Israel recognized them as potential sources of skill. By returning home, the new Argonauts have created economic and professional opportunities for former classmates and for subsequent generations of technical graduates, ultimately reducing the brain drain.

Returning emigrant communities are not replicating Silicon Valley around the world. The emerging regions combine elements of the Silicon Valley industrial system with inherited local institutions and resources. The well-developed institutions that support the Silicon Valley system, including efficient capital markets, property rights enforced by an independent judiciary, regulatory oversight, sophisticated education and research, and technological infrastructure, are rarely all present in peripheral economies. Returning entrepreneurs typically attempt (with varying success) to transfer venture capital finance, merit-based advancement, and corporate transparency to economies with traditions of elite privilege, government control, and widespread corruption. They seek to create team-based corporate cultures with minimal hierarchy in environments that are dominated by family-run or state-owned enterprises.

Nevertheless, returning entrepreneurs have adapted to conditions in their home countries, as attested by the growing presence of technology entrepreneurship in each of these economies. In India, early entrepreneurs relied on private telecommunications facilities and power supplies rather than on the nation's costly and unreliable infrastructure, while in China returning entrepreneurs have learned how to negotiate the Byzantine rules and relationships that regulate private companies. Returning entrepreneurs also have the advantages of access to American institutions: not only do they have the benefit of a graduate education in the United States, but many choose to incorporate their businesses there, establishing headquarters or research labs in Silicon Valley, raising venture capital, retaining professional services, and hiring managerial and technical talent from the United States. Many eventually raise public capital on U.S. stock exchanges. At the same time, in all these countries, cross-regional entrepreneurs and their communities have contributed to the transformation of domestic institutions by advising national governments on legal, regulatory, and capital market reforms, by working with regional governments on improving local infrastructure, universities, research, and training institutions, and by creating forums for information exchange among local companies.

A region's economic trajectory is shaped not only by local institutions but also by the range of technological and market opportunities available at the time it enters global markets. The most successful producers in Israel and Taiwan are those that have identified niches that allow them to differentiate and complement, rather than compete directly with, established producers in Silicon Valley—thus avoiding price and trade wars like those between U.S. and Japanese semiconductor firms in the 1980s. The fast-growing market for wireless communication in Asia has created opportunities for firms in China and India to contribute to the direction of the technology and its applications—even if they do not define the leading edge. Over time, producers in developing regions build independent capabilities and define entirely new markets. Organizational and institutional innovations will also likely emerge from these new centers of technology entrepreneurship, as they did from Japan in an earlier era.

Entrepreneur-led growth, with competitive, sophisticated small- and medium-sized technology producers in high skill regions connecting to and collaborating with counterparts elsewhere, is only one possible future for these formerly peripheral regions. They could forgo the opportunity to upgrade local skills and capabilities, and instead remain suppliers of low-cost labor. China and India have the labor supply to do this for a relatively long time. However, many transnational entrepreneurs have maintained close ties to the technology and markets of Silicon Valley and are constructing firms committed to an alternative, high-value-added trajectory.

ENTREPRENEURSHIP IN THE POSTWAR ECONOMY

Entrepreneurship largely disappeared from scholarly and policy debates in the postwar period. As recently as the 1980s, the general consensus was that the large mass-production corporation represented the optimal way to organize production. Alfred Chandler traced the evolution of the "modern corporation" in the United States to serve a rapidly expanding national market, the adoption of increasingly capital-intensive plant and equipment to maximize throughput and achieve economies of scale, and the related need for control over the sources of supply as well as demand to ensure the ability to amortize these huge initial investments. He also described how the detailed division of labor and delegation of administrative tasks to professional managers within the corporation further lowered unit costs and increased throughput, while a central managerial hierarchy served as "guardian of the organization's centralized knowledge base."[6]

Policymakers in planned as well as market economies attempted to emulate America's unprecedented postwar growth by adopting the organizational model of the twentieth-century modern corporation.

European attempts to master large-scale production began as early as the 1940s, but the process was uneven and often required state intervention at the expense of alternative industrial systems. Scholars have described the hybrid production systems that emerged in Germany, France, and Italy.[7] However, the institutional similarities across these economies, which Alexander Gerschenkron referred to as "late developers," were striking as well—the activist state and leading domestic banks working with giant corporations to mobilize the technology and resources needed for high-volume manufacturing and scale economies.[8]

When microelectronics technologies were commercialized in the 1960s and 1970s, policymakers assumed that the large corporation was the appropriate vehicle for development of this emerging sector. Governments in East Asia as well as Europe sponsored "national champions" in the semiconductor and computer industries that were typically linked closely to national defense, aerospace, and telecommunications agencies. The dominant domestic producers in France, Germany, and Italy, for example, received preferential treatment including public procurement, research funding, and low-cost loans—all justified in the interest of national security and prestige. In Japan and South Korea, the government similarly privileged a few gigantic, diversified electronics and computer producers at the expense of potential competitors, both domestic and foreign, through generous incentives and subsidies. Japan's *keiretsu* and Korea's *chaebol* (large corporate groups or conglomerates) not only became export leaders in technology sectors, but also played dominant roles in the design and operation of financial institutions.

The classic economic model accurately described the postwar period. Long product cycles and stable markets allowed technology firms to shift high-volume production to low-cost locations while maintaining research and development at home. In the 1960s and 1970s, for example, U.S. semiconductor firms relocated their low-skill testing and assembly operations to Southeast Asia, followed in the 1980s by disk drive and other PC producers. The leading computer firms moved high-volume manufacturing to Scotland and Ireland in the 1970s and 1980s to take advantage of low labor costs as well as generous tax incentives and access to European markets. This spatial separation of conception and execution, dubbed by scholars the "new international division of labor" (in contrast with the "old" international division of labor based on natural resource endowments), reflected a simple geographic extension of the Chandlerian corporation.

Entrepreneurship had no role in this system. The large-scale corporation represented the ideal of modern economic progress, while entrepreneurs and small firms were seen as archaic and destined to disappear. Capital and labor markets in Korea and Japan, as in France and Germany, were heavily biased toward established corporations. Immigrant entrepreneurship in the Chinatowns and other ethnic enclaves was regarded as a low-wage survival strategy for recently arrived unskilled workers who lacked opportunities in the mainstream economy. Reliance on ethnic networks to mobilize skill and resources for small-scale enterprises was, in this view, traditional behavior that would be replaced with the passage of time and the inevitable rationalization of economic life.[9]

This interpretation was a self-fulfilling prophecy. As national institutions were organized to promote the development of mass production capabilities, they undermined institutional support for small firms and entrepreneurs, which was in any case less visible. Banks and other financial institutions organized to serve the financial needs of the largest, typically oligopolistic corporations that dominated each sector.[10] Corporations either internalized their sources of supply or relegated suppliers to subordinate and dependent roles that limited their autonomy and capacity to innovate. Government agencies, as well as organized labor, collaborated with the largest firms in a sector to ensure corporate success in exchange for a portion of the surplus. The result was economic progress driven by the state and large corporations. As late as the 1980s, faced with Japan's competitive successes in the automobile and semiconductor industries, business leaders and policymakers in the United States viewed the giant, diversified *keiretsu* as a model for the U.S. technology sector.

Small Survivors

Small firms and entrepreneurs never disappeared, of course. They survived in places that were bypassed by the mass production system, particularly in older urban areas and historic centers of craft and artisan production. In regions like the Third Italy and Germany's Baden Württemberg, local financial and technical institutions as well as social structures evolved to support a dynamic alternative model of decentralized production. Even in the United States, older cities like New York and Philadelphia remained more conducive to entrepreneurship than the new centers of mass production such as Pittsburgh and Detroit.[11] Ironically, just as the mass production system peaked in the 1960s, new technology-based firms were emerging around universities in Boston's Route 128 area and the Santa Clara Valley in California.

These pockets of entrepreneurial dynamism were not limited to the developed world. Taiwan's much-praised "economic miracle," for example, was attributable in part to its networks of small- and medium-sized firms in labor-intensive light industries such as textiles and footwear. This decentralized industrial system was notable for its geographic concentration in central Taiwan, its reliance on indigenous entrepreneurship and local capital, and the ability of its hundreds of specialized firms to coordinate production and adapt quickly to changing world markets. Taiwan's networks of small-scale producers dominated global bicycle and footwear exports throughout the 1970s and 1980s.[12] In the 1970s and 1980s, more than 80 percent of Taiwan's manufacturing establishments employed fewer than twenty workers. The emergence of dynamic small enterprises in Taiwan is linked to mainland emigrants' control of the government and their monopolization of production in upstream industries, which excluded the native-born Taiwanese from both, as well as the success of the land reform in the early 1950s, which created a proliferation of small owner-cultivators.[13]

Economists and policymakers at the time either overlooked these small firms and their networks or regarded them as transitory. Planners in this era promoted large-scale, capital-intensive investments (or "growth poles") to jump-start industrialization in lagging regions of both advanced and developing economies. Many also subsidized the leading national producers and protected the domestic market with the aim of developing indigenous manufacturing capabilities. Scholars focused on the political alliances of multinational corporations, large firms, and the state, which reduced the risks of the large-scale investments required for mass production and also ensured access to the domestic market. In this view, even the most successful newly industrializing countries would remain imitators because the leading technologies would continue to reside in the research and development labs of the large corporations and universities in the core. At best, governments or multinationals might facilitate the adoption of technology developed elsewhere and the incremental mastery of low-end, high-volume manufacturing destined for export markets.[14]

Regional Successes

The unanticipated dynamism of industrial districts from the Third Italy to Silicon Valley revived scholarly interest in regional economies during the 1980s and 1990s. The new economic geographers relied on the notion of increasing returns to scale, and the superior efficiency of the large firm or urban area, to explain agglomeration and regional growth.[15] Yet references to increasing returns or its spatial correlate, external economies, do not by themselves illuminate the mechanisms that underlie productivity gains. At the same time the management strategist Michael Porter, building on his earlier work on national competitiveness, popularized the notion of the industrial "cluster."[16] This concept has the advantage of recognizing the role of institutions, inter-firm relations, entrepreneurship, and innovation (not simply cost reduction) in cluster performance, but it too fails to isolate the distinctive capabilities that differentiate dynamic clusters from their less successful counterparts in the current era.[17]

In the early 1980s Piore and Sabel argued that the resurgence of regional economies was a response to the heightened technological and market uncertainty that was destabilizing the large-scale Chandlerian corporation.[18] However, most business scholars since then have associated the breakup of the vertically integrated firm with the rise of international production networks in which the dominant corporations in the advanced economies responded to intensifying competition by outsourcing production to independent producers in peripheral locations.[19] By viewing developing regions primarily as a source of low-cost labor, this approach overlooks how the increased spatial clustering of production in the periphery has contributed to speed and innovation, hence competitive success—particularly in technology industries that depend on time-to-market and differentiation as well as cost.

The demise of traditional corporate hierarchies and the dynamism of the regional cluster was apparent in Silicon Valley as early as the 1980s. The next section summarizes how institutions and social structures in the region evolved to coordinate a decentralized industrial system in which technological know-how and expertise were diffused across hundreds of specialized start-ups, established companies, universities, colleges, and other training and research institutions—rather than being centralized in one or more corporate R&D labs. The responsiveness and innovative capacity of Silicon Valley's economy became self-reinforcing, as its networks of specialists overwhelmed the adaptive capacity of their established corporate competitors by accelerating the pace of market and technological change.[20]

The breakup of the modern corporation also created new opportunities for entrepreneurship—as well as costs savings—in the periphery. In the concluding section I describe how the new Argonauts built cross-regional communities and transferred both institutions and know-how from Silicon Valley to their home countries. The surprising success of these new entrepreneurial ecosystems in Israel and Taiwan, and more recently in India and China, has transformed global competition once again. While they initially served as sources of low-cost skill, local producers quickly upgraded their own capabilities through learning by doing as well as through collaborations with differently specialized producers in Silicon Valley and elsewhere. The experience of the new Argonauts suggests that regional advantage in today's global economy requires the recombination of both local and distant know-how and expertise to define new solutions and to create new products and industries.

THE TRIUMPH OF THE SILICON VALLEY MODEL

At first the economic dynamism of Silicon Valley was not widely recognized. However, the region flourished during the market volatility and uncertainty that threw America's manufacturing companies and regions into crisis in the 1970s and 1980s. The region's technology upstarts proved more adaptive than their older, vertically integrated counterparts in an environment of fast-changing markets and technological advances. In the 1960s and 1970s, Silicon Valley's semiconductor companies outperformed older East Coast corporations like RCA and Sylvania, and in the 1980s the region's personal computer industry outperformed large, established companies including NCR, Honeywell, and Digital Equipment Corporation. Its decentralized industrial system allowed Silicon Valley as a region to outperform Boston's Route 128 area. Yet as late as the 1980s, scholars and policymakers predicted that Silicon Valley's entrepreneurial small firms would be overwhelmed by the large-scale established technology corporations in the United States and abroad, such as Japan's *keiretsu.*[21]

Wellsprings of Growth

Silicon Valley's growth since the 1970s has been based on entrepreneurship, a deepening division of labor, and open information exchange. The founding myths trace the region's origins to the garage where the Hewlett-Packard Corporation was started, and to the subsequent splintering of the poorly managed Fairchild Semiconductor Company, which spawned several leading successful start-ups including Intel. Frequent references to the Fairchild "family tree" and to the "Fairchildren" and "Grand Fairchildren" suggest the sense of quasi-familial relationships that link these firms and their subsequent spin-offs. This recognition of a common heritage also served to legitimate the high rates of interfirm mobility and information exchange that distinguish the region.

Silicon Valley's founders were young engineers who saw themselves as outsiders experimenting with new technology in a new region, far from the established centers of political and economic power in the United States. They created the shared identities and worldviews that come from ongoing face-to-face interactions, and they developed a common commitment to advancing technology and the region, rather than any individual firm. Lacking the experience or knowledge of how business "should" be organized, these early entrepreneurs also rejected the bureaucracy and hierarchy of established

East Coast corporations. Instead they sought to preserve the energy, the autonomy, and the focus of start-ups even as their companies grew larger. The lack of prior industrialization meant that there were no established interests to block this organizational experimentation. Over time these innovators created an industrial system distinguished by open labor markets, continuous entrepreneurship, and information exchanges both within and between firms—all the direct opposite of the modern corporation with its hierarchical control of information, detailed division of labor and internal labor markets, and corporate secrecy and self-sufficiency.

Supporting Infrastructure

The early emergence in the region of specialized professional service providers, including lawyers, bankers, venture capitalists, and a myriad of consultants well versed in the needs of start-ups and small technology companies, accelerated the expansion of this entrepreneurial system. Venture capital (VC) grew out of informal investments made by the first generation of successful entrepreneurs in their friends' and colleagues' start-ups. Even today, many retiring entrepreneurs recycle their own experiences and networks of contacts by becoming angel or venture capitalists. The Silicon Valley VC industry thus institutionalized the informal sharing of information, experience, contacts, know-how, and advice—along with capital—with new enterprises, and serves as one of many mechanisms for collective learning in the region. A recent analysis of American venture capital–funded start-ups found that the most prolific spawning of startups was by other VC-funded companies in Silicon Valley (in fact, location in the Valley increased the spawning by almost 38 percent), thus supporting the "Fairchild" view of entrepreneurial learning and networks.[22]

The venture capital system created powerful incentives for investors to monitor closely the operations and results achieved by start-ups, since the investors' success depends directly on a relatively small number of high-risk firms in their portfolio. Venture capitalists serve on the board of directors and provide frequent managerial counsel to portfolio companies, while also providing introductions to potential customers, investors, partners, employees, lawyers, commercial bankers, accountants, and others. Silicon Valley firms also pioneered the distribution of stock options to all employees, rather than simply to top management. By linking compensation directly to the success of the firm, stock options can provide a workforce with an incentive to increase shareholder value. The resulting "equity culture" appears to increase the odds of entrepreneurial success while also increasing the ranks of entrepreneurs. Meanwhile, because VC specialists work closely with the firms in which they invest, venture investing remains highly localized. Even today, one-third of all venture capital invested in the United States goes to firms based in Silicon Valley (and this pool is more than double the amount invested in New England, the second-ranked region).

The region's law firms likewise have developed deep expertise in issues relevant to technology start-ups: intellectual property rights, initial public offerings, technology licensing, venture financing, and trade law. Local accountants, human resources firms, early-stage investors, investment bankers, marketing and public relations firms, and a diverse and changing community of consultants also develop business and reputations by networking, providing referrals, and collaborating with one another.

The founders of the Silicon Valley Bank, which was started in 1983 by a team of former Bank of America managers, recognized the coordinating role of professional service providers in this entrepreneurial ecosystem: "These new companies were struggling to survive in an economic landscape that didn't fit their needs. Support businesses for entrepreneurs were not available, and most traditional business and professional service providers weren't interested in risky pre-revenue, pre-profit ventures." The bank describes its competitive advantage as its credibility and reputation as "an integral member and a trusted partner in the region's networking infrastructure . . . providing relationships, networking, and referrals" in an "economic ecosystem that has evolved since the early 1980s and consists of partnerships, alliances, and interdependencies among: (1) venture capital firms, (2) professional service providers, and (3) entrepreneurial companies and their managers." According to its Web site, by 2000 the Silicon Valley Bank was a major presence in the region and had twenty-six U.S. branch offices, some $4 billion in assets, and a thousand employees.[23]

These specialized professional service providers give Silicon Valley entrepreneurs an advantage relative to those in older industrial regions dominated by established, and often more conservative, legal and financial communities. It has been said that an entrepreneur can move from initial concept to incorporation, financing, and growth more rapidly in Silicon Valley than anywhere else in the world. In the words of repeat entrepreneur Min Zhu, cofounder of WebEx:

> Where do you learn to run a company? There are so many things that can go wrong in a start-up; when you begin you don't know more than 10 percent of what might go wrong. The advantage of Silicon Valley is that you can fail and learn and try again. It's the only place where you can screw up once and try again.
>
> You also can't grow a company this fast anywhere else. We bootstrapped WebEx ourselves with no salary while we developed our product, hired thirty engineers with low salaries and high stock options. (They went out and consulted on the side to make money.) It took about eighteen months to create the product, and then we sold it. We raised $7 million from an angel investor, and then took more traditional venture capital. In the first round we raised $28

million and in the second we raised another $25 million. Less than two years later we took the company public.[24]

DECENTRALIZATION AND MARKET FRAGMENTATION

The Silicon Valley system supports far more competing experiments with new technologies, products, and applications within a locality than would be possible within the boundaries of a large corporation. This proliferation of alternative technological pathways and opportunities was evident as early as the late 1960s. According to Intel cofounder Gordon Moore, "Integrated circuits, MOS transistors, and the like proved too rich a vein for a company the size of Fairchild to mine, resulting in what came to be known as the 'Silicon Valley effect.' At least one new company coalesced around and tried to exploit each new invention or discovery that came out of the lab."[25] This proliferation of parallel experiments ensures both competition and redundancy in the economy, further strengthening the entire system.

The uncoordinated but continuous emergence of specialist firms pursuing new technical and market opportunities supports the ongoing fragmentation of production as well as markets. An independent semiconductor equipment and materials industry emerged in the region in the 1960s and 1970s as engineers left the established companies to start specialized firms that manufactured the capital goods the industry demanded—diffusion ovens, step and repeat cameras, testers, and materials and components such as photo masks, testing jigs, and specialized chemicals. This independent equipment and materials sector in turn encouraged the formation of new semiconductor firms by freeing individual producers from the expense of developing capital equipment internally, and by spreading the costs and risks of equipment development over a wider community. It also reinforced the tendency toward industrial localization, since most of the specialized equipment and inputs were not available elsewhere at the time.

A handful of these entrepreneurs founded the Semiconductor Equipment and Materials Institute in the 1970s as a forum for local producers to share information. One of SEMI's most significant early roles was in setting the technical standards for the emerging industry, without which specialization would not have been possible. The Institute also provides industry and business data, organizes trade shows, and facilitates technological and information exchange among the enterprises in this highly fragmented industry. The organization—now named simply SEMI—has 2,200 members and offices in Shanghai, Beijing, Taiwan, South Korea, Japan, Singapore, Europe, and Russia in addition to the United States.

The Silicon Valley technology industries, from semiconductors to personal computers and their components to telecommunications and the Internet, have all evolved through this process of specialization and unbundling; in turn, they have contributed to the formation of unanticipated new markets and cross-fertilizations. The fragmentation contributes to the diversification and resilience of the local economy while also enhancing opportunities for local innovation, since specialized components can be recombined in a multiplicity of different ways to create new applications—what the economist Allyn Young referred to as the increasing "roundaboutness" of production.[26] Of course, it is possible to specialize without innovating and to innovate without changing the division of labor. However, the deepening division of labor can enhance the innovative capacity of a community. As expanding opportunities for experimentation generate ideas, these ideas are in turn combined to make new ideas, and so on in a dynamic and self-generating process.

Specialization can thus increase both innovation and economic growth. Most Silicon Valley entrepreneurs do not simply replicate existing products at a lower cost, grinding out so-called "me too" products; rather, they define new markets, technologies, processes, and designs that offer new capabilities, features, and qualities. This combination of specialization and diversification ensures that the difficulties of a single firm cannot destabilize its industry, and even the failure of an industry cannot destabilize the entire region.

The high rate of failure as well as of new firm formation is a crucial source of collective learning in Silicon Valley. The rapid turnover in industrial leadership in the region reflects this process of creative destruction. The dominant companies in Silicon Valley have changed almost completely since 1980. Half of the firms that ranked among the forty largest high-tech firms in 1982 no longer existed in 2002; and only four firms from the 1982 list were still among the top forty in 2002. Likewise, more than half of the top forty firms in 2002 had not even been founded in 1982.[27] This turbulent system appears chaotic, even wasteful, yet the region—if not all of its firms—remains a leader in technological and organizational innovation after more than four decades.

One indicator of the region's sustained innovative capacity is the number of patents. Silicon Valley inventors have gained a growing share of patents issued by the U.S. Patent and Trademark Office in recent years. Although this trend reflects today's greater emphasis on protecting intellectual property, Silicon Valley has increased its total proportion of U.S. patents. The region's inventors were recipients of 4 percent of all U.S. patents awarded in 1993; by 2003, their share had grown to 10 percent (and they received close to 45 percent of all patents awarded in the state of California). Silicon Valley's inventors have accelerated the rate of patenting as well, from 103 patents per 100,000 residents in 1993 to 371 in 2003. The national average for this entire period is fewer than 50 patents per 100,000 residents.[28]

Creating a Local Technical Community

In traditional mass production corporations, technologists remained inside the firm, typically in the R&D laboratory or engineering division. In Silicon Valley, by contrast, technical communities transcend companies and other institutions including universities, colleges, and research labs. The early collaboration between the once small and undistinguished Stanford University and the region's fledgling electronics enterprises created a model of how local institutions could adapt to the changing needs of business in the region. Economic historians have documented the mutually beneficial co-evolution of the two, "gradually adjusting to each other and their common competitive environment. Each helped the other discover and exploit new niches in science and technology . . . fostering horizontal integration and collective learning throughout the region."[29] This process of mutual adaptation and bootstrapping was repeated across the region in relations between entrepreneurs and established companies, and between firms and community colleges, research institutions, and professional service providers; the process also showed up in the relations among subsequent generations of entrepreneurial businesses. The result was a blurring of traditional boundaries between finance, law, business, education, and local government.

Engineers in the postwar period traditionally assumed that they would spend their entire career in a large corporate or university laboratory like Bell Laboratories or MIT's Lincoln Lab. In Silicon Valley, however, engineers quickly came to see themselves as potential managers or entrepreneurs in a fast-changing market. Many developed entrepreneurial aspirations as a way to pursue autonomy as well as technological or commercial opportunities—and few saw their careers in a single company. Over time, as local engineers became managers or investors and as university researchers became entrepreneurs or consultants, they developed collective identities as part of the region's industrial and technical community rather than as employees of any particular firm, institution, or even sector.

The paradox of the Silicon Valley economy is that a key source of value, the know-how of individual workers, diffuses rapidly between local firms because of the region's open, high-velocity labor markets. Firms invest heavily in intellectual property and compete fiercely with one another, but at the same time, rapid interfirm mobility among skilled employees ensures the widespread diffusion of the know-how that facilitates innovation, adaptation, and growth. The authors of a recent analysis of employee mobility in the U.S. computer industry found substantially more job hopping in Silicon Valley than in other concentrations of computer production, which they associate with "ubiquitous knowledge spillovers."[30]

This is not to suggest that Silicon Valley's competitive advantage derives from employees' transferring trade secrets to competing firms. In fact, allowing companies in Silicon Valley to misappropriate trade secrets from their competitors would provide a powerful disincentive to invest in innovation. Rather, the Silicon Valley experience suggests that it is the diffusion of a wide range of know-how of individual employees, not their employers' trade secrets, that accelerates the pace of innovation and growth; it does so without undermining the incentive to innovate.

Start-ups in Silicon Valley rarely compete directly with established producers. New companies typically differentiate their products or services by defining a new market niche, process, or technology which they hope will provide a distinctive competitive edge. This means that an employee can work on a competing product for a new employer without copying the former employer's product, and that differently specialized partners regularly share information about long-term product and technological plans in the interest of joint learning.

The regular movement of engineers between Silicon Valley firms and the corresponding diffusion of know-how are now widely accepted within the technical community, as are the benefits of practices such as investing in former employees' start-ups, collaborating with research and educational institutes, and participating in standard-setting activities. These practices have been institutionalized in the region's high-velocity labor markets in part because California provides legal protection for employees who move between firms. California law, unlike that in Massachusetts and many other states, all but ignores non-compete agreements. As a result, the region's employees have been free to move to competing firms and to work on related products as long as they do not disclose or use the proprietary information of their former employer.[31]

The shared educational and professional backgrounds of local engineers and entrepreneurs provide a common language and formative experience. Departing employees typically maintain good relations with former employers and colleagues, recognizing that they may become future customers or partners. In short, careers in Silicon Valley overlap and diverge repeatedly. Veteran venture capitalist John Shoch concludes: "They say that there are six degrees of separation between all of us. In Silicon Valley, it is closer to two degrees of separation."[32]

A variety of more or less formal professional, technical, and business organizations, alumni associations, standard-setting forums, and informal hobbyist groups continue to emerge in the region, providing opportunities for face-to-face interaction and exchange across firm, industry, or sector boundaries. Some of these associations are short-lived, associated with narrow technical and market opportunities, while others have remained part of the regional landscape for decades. These networks accelerate information flows, new firm formation, and the formation of partnerships—even among competitors. They facilitate the rapid identification of market opportunities, sources of funding, and team members, as well as suppliers, customers, and partners. These networks also accelerate the spread of reputations, both positive and negative, and they socialize newcomers to entrepreneurship by exposing them to successful role models as well as to potential mentors and promising business opportunities.

All these features of the technical community enhance its performance, while also significantly reducing the risk associated with new firm formation and even failure. Entrepreneurs in the region can quickly build teams and grow new companies by drawing on a rich skill base that includes diverse scientific and technological capabilities, a depth of professional and managerial experience, and networks in the technology community. Equally important, Silicon Valley remains an open community, and one in which professional reputations are carefully managed. The respect for technical merit facilitates recruitment of qualified outsiders, even those of different ethnic or racial backgrounds. The orientation toward commercial success encourages engineers to become entrepreneurs, while the emphasis on team-based development ensures loyalty for the duration of a project. The "small world" nature of the community results in the responsiveness and intensity of communication that come from personal relationships. In contrast with the internal career ladders of the modern corporation, the open labor markets and entrepreneurial infrastructure of Silicon Valley make firms that become closed to outside resources or ideas unlikely to survive.

Silicon Valley Meets the Modern Corporation

The rising costs of new product development and the accelerating pace of innovation in electronics and computing in the 1970s and 1980s meant that Silicon Valley start-ups could not compete in all markets with the established, vertically integrated systems corporations such as Honeywell, IBM, Control Data, and Digital Equipment Corporation. (Those start-up companies that tried to integrate vertically, for example from semiconductors into products like digital watches, were forced to retreat.) Local start-ups instead focused on a particular component or subsystem, while relying on a diversified range of other local specialists to provide other components and final system integration.

The opening up of the hardware and application software systems of the IBM personal computer (PC) in the early 1980s, in response to the ongoing pressure of antitrust actions by the U.S. Department of Justice, was crucial to this shift. The System/360 family of mainframe computers was designed with a common operating system and standardized interfaces that allowed for modular and upgradeable processors and peripherals. This enabled suppliers to manufacture standard plug-compatible components for the system, which facilitated lower development expenses and production costs while allowing entry of external peripheral producers. The IBM PC, which was introduced in 1981, was considered innovative in part because it was based entirely on outsourced components. IBM opened both its hardware components system (CPUs, motherboards, mice, disk drives, printers, and the rest) and its software (operating system as well as application software). Even internal component divisions were required to compete to supply IBM for these contracts. The PC created a mass market for personal computers and triggered the proliferation of new computer systems firms as well as literally thousands of new producers of a diverse range of components, peripherals, and applications—including two of the most valuable and dominant IT franchises: Intel and Microsoft.

Producers in Silicon Valley had a tremendous advantage in the new competitive environment. The region was already home to a sizable community of PC hobbyists and software programmers, along with early computer firms like Apple and Commodore. The local expertise and experience base in semiconductor design and manufacturing was unsurpassed globally; the IBM research facility in San Jose was at the leading edge of disk drive technology; Palo Alto–based Xerox PARC was a pioneer of leading innovations in PCs (even though it famously failed to commercialize them); and both Stanford University and U.C. Berkeley were centers of research in electronic engineering and computer science. These research institutions not only provided top-tier talent for local industry but also continued to attract the best engineers from around the country—and the world. By the late 1980s, Silicon Valley–based producers were introducing innovative new components and peripherals faster than their more integrated competitors from elsewhere in the United States and abroad. The dynamism of the region's specialists was evident in new generations of specialized integrated circuits and microprocessors, disk drives and printers, software and networking hardware.

The traditional vertically integrated systems corporation gave way under these competitive pressures to a fragmented industry structure with increasing firm specialization. The number of actors in the computer industry increased dramatically, and competition within most technology segments intensified greatly. Today independent companies produce all of the components that were once internalized within a single corporation—from applications software, operating systems, and computers to microprocessors and other components. Within each of these segments there is, in turn, increasing specialization of production and a deepening division of labor.

Fragmentation of the Semiconductor Industry

The semiconductor industry illustrates how the deepening division of labor opens up opportunities for new companies by lowering costs and expanding markets. In the 1960s and 1970s, vertically integrated corporations like IBM and NEC performed all the tasks required to produce an electronic system, from the initial system architecture and integrated circuit (IC) design, to IC manufacturing and assembly, to final system assembly. In the 1970s a new generation of integrated device manufacturers (IDMs) like Intel and National Semiconductor focused exclusively on IC design, manufacturing, and assembly, leaving system architecture and integration to other firms. An independent semiconductor equipment manufacturing and materials industry emerged in Silicon Valley at the same time.

In the 1980s, Taiwan Semiconductor Manufacturing Corporation introduced the stand-alone IC foundry to manufacture chips designed elsewhere, which facilitated further unbundling of IC manufacturing from both system and IC design and from final assembly and testing. The late 1990s saw the emergence of independent new suppliers of intellectual property (IP) modules that go into chip design, as well as specialists in process R&D providing services to the IC foundries. The new specialists can be found in Israel, India, and China as well as in the United States and Japan.

The continuing fragmentation of the semiconductor industry has opened up a wide range of opportunities in the areas of chip materials (for example, copper interconnects, silicon on insulator, silicon germanium, strained silicon), new architectures and circuits (system on chip, magneto resistive RAM, double gate transistors, carbon nanotube transistors, biological and molecular self-assembly), manufacturing or fabrication initiatives (new processing chemistries and integrated metrology), and packaging processes (such as flip chip, wafer scale packaging, 3D and system in a package).

These innovations in semiconductor design and fabrication have supported a proliferation of new applications and opened up new markets for specialist producers.[33] Companies that failed to recognize the competitive advantages of specialization during the 1990s, mainly in Japan, South Korea, and parts of Europe, have fallen behind in world markets. For example, the Japanese producers NEC and Fujitsu lost ground as they distanced themselves from the open-system PC trends in the 1980s and 1990s, instead continuing to rely on closed, proprietary operating systems and cutting themselves off from the accelerating pace of innovation in custom ICs, application software, and other components occurring elsewhere in the world.[34] The problems of South Korea's semiconductor industry in the 1990s similarly derived from the vertical integration and inflexibility of a structure organized around giant export-oriented *chaebol* engaged in commodity DRAM production.[35]

Open Networks, Collective Innovation

Few in the semiconductor industry would have anticipated the continuing specialization of production over the past two decades, or the extent to which collaboration among specialists has resulted in products or processes that none of the participants anticipated. The coordinating functions of the corporate hierarchy (and even the lead company in a supply chain) are being replaced by shifting collaboration between specialist producers that jointly design and produce a range of components and subsystems. This decentralization of production depends upon common technical and communication standards and interfaces that are defined either by dominant firms, or increasingly by large-scale consortia of specialized producers.

Entrepreneurial success in this environment depends upon access to robust information exchange networks and the ability to efficiently locate other specialist collaborators. This explains why a disproportionate number of new firms start in regions with other sophisticated specialist producers and concentrations of related expertise, in spite of the high cost of doing so. As Silicon Valley's entrepreneurs innovate in increasingly specialized niche markets, intense communications ensure the speedy, often unanticipated, recombination of these specialized components into changing end products. Location in an environment like this enhances a firm's ability to identify and respond quickly to potential market niches. Firms in such a system recognize the value of collaboration in a process of mutually beneficial bootstrapping, which some refer to as "growing up together."

Specialization not only frees designers to focus, innovate, and grow their markets; it also allows them to share the risks of exploring new technological and market opportunities with other specialists.

In markets plagued by uncertainty and very short product cycles, customers and suppliers often collaborate simply in order to keep up with each other. For example, the rapid transition in the 1990s from standard mobile phone handsets to widely varied designer mobile phones, and to mobile phones with PDAs and wireless Internet access and digital cameras or mobile multimedia capabilities, was not achieved through the design of leading companies. Rather, it was the result of concurrent design and production among differently specialized firms located mainly in Taiwan and China.[36]

While partners need not be geographically proximate to collaborate, collocation is a tremendous asset, as are a shared language, culture, and worldview. Stan Shih, founder and CEO of Taiwan's Acer, summarizes the innovative benefits of location in Taiwan's Hsinchu Science Park or Silicon Valley:

> Industry clustering speeds up the pace of innovation. Once . . . clustering is established, working within its disintegrated structure can allow an individual business to concentrate its capabilities on a certain task and share risks with other companies. If there are risks associated with a larger initiative, tasks can be appropriately allocated and coordinated among upstream, midstream, and downstream satellite businesses; and if by chance the direction is wrong, everyone involved can communicate and the plan [can be] adjusted quickly. Even if some losses are incurred, they are shared by all the parties involved, and the loss for individual companies is minimized.[37]

ENTREPRENEURSHIP IN THE PERIPHERY

The social structure of a technical community appears essential to the organization of production at the global as well as the local level. In the old industrial model, when technical communities were primarily inside corporations, the company was the primary vehicle for the creation and transfer of knowledge.[38] In regions like Silicon Valley the technical community transcends firm boundaries, and such tacit knowledge is often transferred through informal communications or the movement of individuals between firms. As producers innovate in increasingly narrow niche markets, they collaborate with customers, suppliers, and partners to learn about, and respond to, a fast-changing, competitive environment. Timely information and industry-specific know-how accumulate in local universities, venture capital networks, specialist consultants and service providers, and a variety of technical, professional, and alumni associations. The knowledge that once resided within the corporation is transferred through intense informal communications and the frequent movement of individuals between firms, as well as interfirm collaboration. This suggests that the stand-alone modern corporation of the twentieth century may no longer be essential to the transfer of technology or skill across national borders; an international technology community provides an alternative and potentially more flexible and responsive mechanism.

As recently as the 1970s, only the world's largest corporations had the resources and capabilities to grow internationally, and they did so by establishing manufacturing or marketing subsidiaries overseas. Now digital information and communications technologies mean that even the smallest firms and start-ups can access distant skill and know-how. It became increasingly clear during the 1990s that large corporations no longer controlled the pace of technological change and new product introduction. In the words of HP executive Bill Ting:

> Technology today flows in all directions. The world is incredibly interdependent when it comes to high-technology industries; we are sourcing product assembly, design, manufacturing, and procurement all over Asia. Hewlett-Packard has an entire division dedi-

cated to identifying local companies around the world that are capable of supplying components and sub-assemblies on an OEM/ODM basis.[39]

As managers recognize that innovation is no longer the exclusive province of internal R&D labs, or even of the advanced industrial economies, they increasingly turn to external sources of ideas and information.[40] Large corporations often seek to capture the benefits of entrepreneurship by spinning off divisions, creating internal start-ups, or launching venture capital arms to invest in outside start-ups. Today companies are more likely to form partnerships to develop technology jointly with customers or suppliers, pay to acquire external technology, or set up R&D labs in new locations around the world. These tactics enable larger corporations to participate in the entrepreneurial activity that characterizes regions like Silicon Valley. Foreign-born managers in these companies often play a central role in bridging distant centers of technology and entrepreneurship. Many of them are early and effective champions for establishing a company's overseas operations in their home countries, acting as pioneers in relatively uncharted terrain.

A cross-regional technical community allows distant producers to specialize and collaborate to upgrade their capabilities, particularly when the collaborations require intense communications and joint problem solving. The trust and local knowledge that exist within technical communities, even those that span continents, provide a competitive advantage in an environment where success depends on being very fast to market with new products. Rather than competing for a relatively fixed market, these specialists are jointly growing their markets by introducing new products, services, and applications. As a result, while the relationship between producers in distant regions deepens over time, they can remain complementary and mutually beneficial rather than zero-sum.

Entrepreneurship is risky by definition. Even in the best circumstances, a majority of start-ups fail or grow very slowly. However, regions like Taiwan and Israel have created systems of entrepreneurial growth that encourage start-ups and increase the likelihood of business success. Entrepreneurs actively reshape the local environment as they grow their firms by supporting one another and by working to influence policy. Success reinforces success in these regions, as spinoffs from successful companies define new markets and technological pathways while also providing training grounds and role models for subsequent entrepreneurs.

The entrepreneurial ecosystem is still in its formative stages in the technology regions of India and China. These regions have seen important early entrepreneurial successes, and both have large, productive, low-cost, and technically skilled workforces. However, few U.S.-educated emigrants returned to either India or China during the 1980s and 1990s. The recent technology recession triggered an upsurge in cross-regional entrepreneurship that, along with the emergence of a second generation of successful start-ups, should significantly strengthen the local environment for entrepreneurship in both countries. While it may be too early to define the trajectories of the regions in these large, complex political economies, the cross-regional entrepreneurs and their communities have pursued very different strategies: returning Chinese entrepreneurs have focused primarily on developing products to serve the domestic market, while their Indian counterparts are oriented toward providing software and other services for export.

This suggests that localization is a complement to, not an adversary of, economic globalization. Globalization is increasingly a process of simultaneously designing and integrating specialized components through cross-border collaboration. Companies specialize in order to become global, and their specialization in turn allows them to add value as collaborators. The best environments for breeding specialist firms are the industrial systems of places like Silicon Valley.

Just as the social structures and institutions within these regions encourage entrepreneurial bootstrapping, so the creation of a cross-regional technical community facilitates collaborations with producers in distant regions and a process of reciprocal industrial upgrading.

The recent successes described here—the experiences of Israel, Taiwan, China, and India—suggest that new centers of technology and entrepreneurship cannot be created in isolation. Rather, they require ongoing connections to the U.S. market—often through integration into Silicon Valley's technical community. These cases suggest as well that regions seeking to compete in global technology markets should devote resources to expanding education, adapting the local infrastructure, and building ties to customers and sources of technology in Silicon Valley rather than simply seeking to lure foreign investment or to create an industrial cluster in isolation.

2 ❖ Learning the Silicon Valley System

Immigrant entrepreneurs are both the product of Silicon Valley and the most important carriers of its industrial system. As the region's technology firms grew in the postwar period, they attracted talent from all over the United States, typically directly out of graduate programs in American universities. In the 1970s and 1980s employers began hiring immigrants in growing numbers because they represented a growing portion of university engineering graduates. By 1990 foreign-born professionals, a majority from Asia, accounted for more than a third of Silicon Valley's high-skilled technology workforce, yet their presence was unrecognized outside the region. It was only later in the decade as these immigrants became entrepreneurs that they gained national attention. By 2000 scores of Indian and Chinese engineers, including Jerry Yang, co-founder of Yahoo, Pehong Chen, founder of Broadvision, Vinod Khosla, co-founder of Sun Microsystems, and Sabeer Bhatia, founder of Hotmail, had achieved international recognition together with substantial wealth as a result of entrepreneurial success in Silicon Valley.

REWEAVING THE SOCIAL NETWORK

Like their U.S.-born counterparts, the early immigrant engineers in Silicon Valley began their careers in the "golden rice bowls"—the research laboratories of established corporations like IBM or Bell Labs. In time, many pursued entrepreneurial opportunities of the kind that had come to characterize Silicon Valley. Feeling like outsiders to the old-boy networks created by the region's native-born engineers, the immigrant Indian and Chinese engineers organized their own social and professional networks. They found one another socially first because of shared educational and professional backgrounds, as well as common culture, language, and history. Over time they adapted the networks to professional ends, providing first-generation immigrants with access to role models, contacts, advice, funding, and the local market knowledge needed to identify partners and business opportunities.

The first immigrant networks in Silicon Valley were largely informal gatherings of Iranian, Israeli, and French professionals in the 1970s, followed by the first formal Chinese and Indian associations in the 1980s. This fabric of ethnic professional and technical associations deepened and diversified rapidly in the 1990s, allowing subsequent cohorts of immigrants to integrate into Silicon Valley's labor markets and social networks. These associations created forums for the continuous information exchange and communication that support successful entrepreneurship. Perhaps most important, these ethnic professional networks helped first-generation immigrants overcome the barriers to promotion that many faced in the region's established companies. A distinguishing feature of the local economy is the opportunity for career advancement provided by entrepreneurial companies.

By the late 1990s, Indian and Chinese immigrants in the region had mobilized thousands of their colleagues into active associations—networks that helped to create shared identities and trust. Like all immigrant associations, they helped newcomers to assimilate quickly. Not surprisingly, the rate of immigrant entrepreneurship in Silicon Valley increased dramatically in the second half of the 1990s, along with growth of the region's foreign-born workforce. The quip among semiconductor engineers was: "When we say that Silicon Valley is built on ICs, we don't mean integrated circuits—we mean Indians and Chinese."

The standard image of an immigrant entrepreneur is that of a low-skilled foreigner in a marginal retail or sweatshop manufacturing business located in an ethnic enclave, like a Chinatown. Silicon Valley's new Argonauts, by contrast, typically hold postgraduate degrees and are a growing presence in the most technologically dynamic and global sectors of the world economy. Like their less-educated counterparts, however, these foreign-born professionals pursue ethnic strategies to enhance their entrepreneurial opportunities. The region's most successful Chinese and Indian entrepreneurs appear to rely on such ethnic resources while simultaneously integrating into the mainstream technology economy.

THE BRAIN DRAIN

The growth of regions like Silicon Valley fueled the exodus of talented young adults from poor developing countries to the United States. During the years following World War II, American universities attracted the most promising science and engineering graduate students from around the world. By the early 1990s, 40 percent of the science and engineering doctorates in U.S. universities were awarded to foreign-born students.[1] Students from East and South Asia (65 percent of the total) dominated, with the majority coming from mainland China (22,448), Taiwan (10,926), India (9,981), and South Korea (9,805). Entire graduating classes from the elite engineering program at National Taiwan University, for example, came to the United States in the 1980s, as did a majority of engineering and computer science graduates from the prestigious Indian Institutes of Technology. Technical universities from smaller countries like Ireland and Israel also report large proportions of graduates leaving to study in the United States, although their numbers are too small to show up in the aggregate data.

Postwar reforms of the U.S. immigration system dramatically expanded the opportunities for foreign-born students to remain in the United States after graduation. The most important reform was the creation by the Hart-Celler Act in 1965 of a new category of visas for those whose skills were in short supply, and the simultaneous lifting of national quotas on immigration.[2] Subsequent reforms in 1980 and 1990 expanded the number of foreign engineers admitted to the United States on the basis of scarce skills.

As a result, when Silicon Valley's technology companies created some 150,000 jobs between 1975 and 1990, the region's foreign-born population more than doubled. By 1990, one-third of the scientists and engineers working in the region's high-tech industries were foreign-born, with Chinese (from Taiwan, Hong Kong, and mainland China) and Indians accounting for two-thirds of the total. In the United States as a whole at this time, only 15 percent of the scientists and engineers were foreign-born.[3] Skilled immigration to the United States accelerated again in the 1990s following the near tripling of visas for scarce occupational skills—from 54,000 to 140,000—by the 1990 Immigration and Nationality Act.

Taiwanese immigrants with graduate degrees from U.S. science and engineering programs flocked to Silicon Valley in the 1970s and 1980s. They were joined in the 1990s by still larger numbers of mainland Chinese and Indian immigrants. This increase reflects the lifting of Chinese restrictions on overseas travel and study that began in the early 1980s and was amplified by the events of Tiananmen Square in 1989. By the late 1990s almost a third of the graduates of the elite Peking University, for example, close to a thousand students a year, left China to study overseas, mainly in the United States. A comparable proportion of the engineering graduates of India's prestigious Indian Institutes of Technology emigrated to the United States in the 1990s.

IC VALLEY

Silicon Valley attracted skilled engineers from all over the world, but especially from India and China. Unlike many other developing countries, India and China educate their most talented youth to standards that ensure admission to U.S. universities.[4] The size of mainland China's population makes it a huge player; it dominated high-skilled immigration to the San Francisco Bay Area between 1985 and 2000, with a total of 51,916 professional and technical workers admitted, compared to 26,765 from India and 18,651 from Taiwan (Figure 2.1).

The region's technology industry is also home to large numbers of workers from the Philippines and Vietnam, although most of them occupy lower-level technical positions that do not require graduate degrees. The most talented young people from advanced economies like Japan and France are either less likely to come to the United States for higher education or more likely to return home directly after graduation. Even South Koreans, who received science and engineering doctorates in the United States at comparable rates to those of Indians and Chinese in the 1980s and 1990s, could return home to secure jobs in large, prestigious companies. In the early 1990s only one-third of Korean students planned to stay in the United States after graduation, compared to 94 percent of Chinese students and 86 percent of Indians. As a result, the Korean professional community in Silicon Valley today remains significantly smaller than the Indian and Chinese communities.[5]

The 2000 U.S. Census showed a dramatic increase in the number of Asian-born scientists and engineers working in Silicon Valley, par-

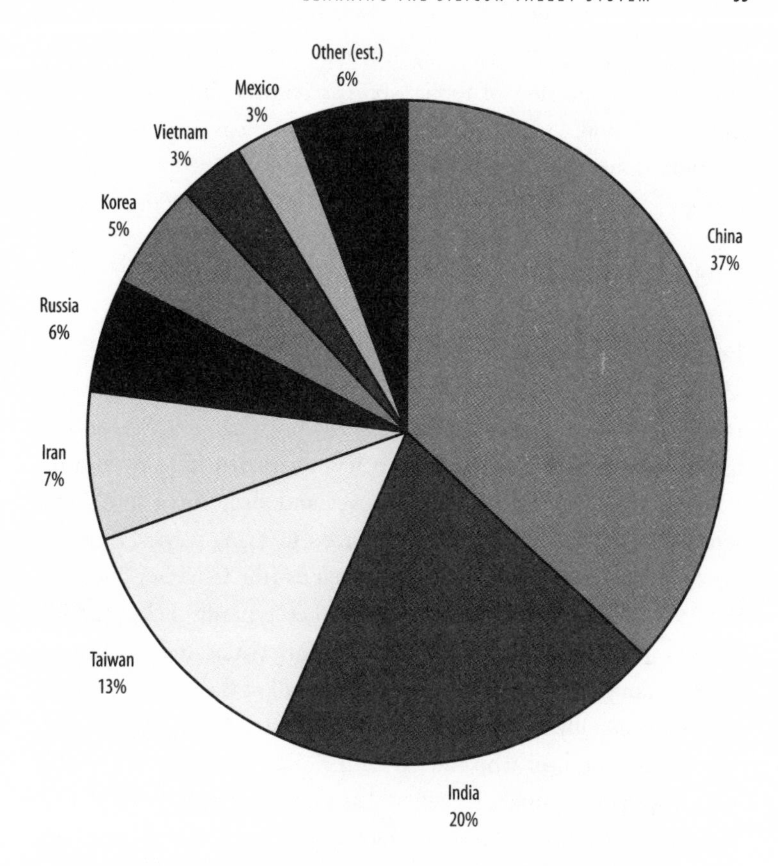

2.1. Professional and technical immigration to the San Francisco Bay Area, 1985–2000.

ticularly Indians, whose numbers grew more than five-fold to 13 percent of the total (up from 4 percent in 1990), followed by mainland Chinese with 9 percent and Taiwan-born Chinese with 4 percent. According to the Public Use Microdata (PUMS) furnished by the census, by the year 2000 first-generation immigrants accounted for a majority (53 percent, compared to 30 percent in 1990) of the scientists and engineers working in Silicon Valley's technology industries. This proportion of foreign-born scientists and engineers was more than double that of other U.S. technology regions such as Boston's Route 128, Seattle, or Austin.

MOBILIZING ETHNIC PROFESSIONAL NETWORKS

Lester Lee, a native of Szechwan who moved to Silicon Valley in 1958, describes the feeling of being an outsider that was common for the region's early immigrants. "When I first came to Silicon Valley," he remembers, "there were so few of us that if I saw another Chinese on the street I would go over and shake his hand." This sense of being an outsider was reinforced in many ways. Lee notes, for example, that "nobody wanted to sell us [the Chinese] houses in the 1960s." Although immigrants like Lee typically held graduate degrees in engineering from American universities and worked for mainstream technology companies, almost all of them felt personally and professionally isolated in a world dominated by Caucasians. David Lee (an engineer who started with Xerox; not related to Lester) adds that many immigrants believed as well that employers saw them as "good work horses, and not race horses" or "solid technicians, rather than promising managers."[6]

A 1991 survey of Asian immigrant professionals in the region found, for example, that two-thirds of those working in the private sector believed that advancement to managerial positions was limited by race. Moreover, these concerns increased significantly with the age and experience of the respondents. This perception is consistent with the finding that in the technology industry, at least, Chinese and Indians remained more concentrated in professional as opposed to managerial positions than their Caucasian counterparts, despite comparable or superior levels of education. It is notable, however, that those surveyed attributed these limitations less to overt "racial prejudice and stereotypes" than to the perception of an "old boys' network that excludes Asians" and the "lack of role models."[7]

These perceptions have a clear basis: the superior educational attainment of Silicon Valley's Asian immigrants is only partially reflected in occupational status. In 1990, Indians and Chinese were disproportionately concentrated in professional (technical and engineering) rather than managerial occupations when compared to their Caucasian counterparts. The relatively lower representation of Chinese and Indians in managerial positions could be due to several factors—biases favoring technical as opposed to business education, or the linguistic and cultural difficulties of many new immigrants. It could also reflect more subtle forms of discrimination or institutional barriers to mobility—the well-known glass ceiling, but based on race rather than sex.

Income data provide little support for the glass ceiling hypothesis. There is no statistically significant difference between the earnings of Chinese and Indians in managerial, professional, and technical occupations and their white counterparts.[8] This is consistent with the findings of other researchers, who document greater disparities in managerial representation and upward mobility than in wages between Asian and Caucasian engineers with comparable skills and education.[9]

Chinese Pioneers and New Organizations

Immigrants responded to the sense of exclusion from established business networks and social structures in two ways. Many started their own businesses. Lester Lee, for example, became the region's first prominent Chinese technology entrepreneur when he left Ampex in 1970 to start a company called Recortec. Three years later, David Lee left Xerox to start Qume after the company hired a less experienced outsider as his boss. He raised start-up capital from the mainstream venture capital community, but only on the condition that he hire a non-Asian president for his company. Similarly, David Lam left Hewlett-Packard in 1979 after being passed over for a promotion and started a semiconductor equipment manufacturing business called Lam Research, which is now a publicly traded company with $1.3 billion in sales. Not surprisingly, these three men became role models for subsequent generations of Chinese entrepreneurs.

Like generations of less fortunate immigrants before them, immigrants to Silicon Valley responded to the sense of exclusion from the region's mainstream technology community by organizing professional associations. Early gatherings were often social celebrations of holidays or even family events. Over time the gatherings turned to professional purposes, providing resources and role models to assist individuals within the technology community. Lester Lee, David Lee, and David Lam, for example, formed a local branch of the Chinese Institute of Engineers (CIE) in 1979 to promote better communication and organization among the region's Chinese engineers. The Bay Area chapter of CIE quickly became the largest in the country; today CIE has some one thousand members in the Bay Area and is regarded by old-timers as the "grandfather" of Silicon Valley's Chinese technology organizations.

Connecting Indian Immigrants

Indian engineers in Silicon Valley faced similar challenges. One old-timer asks: "Why do you think there are so many Indian entrepreneurs in Silicon Valley? Because they know that sooner or later they will be held back." When a young Intel engineer and his three Indian roommates started the Silicon Valley Indian Professionals Association (SIPA) in 1987, their goal was to provide a meeting place for

Indian professionals to share their common concerns. In spite of their mastery of English, which distinguished them from most of their Chinese counterparts, they too were concerned about limits on their opportunities for professional advancement. According to SIPA founder Prakash Chandra, "many Indians didn't see a career path beyond what they were doing." The early SIPA meetings were therefore focused on individual career strategies as well as on the nuts and bolts of the technology industry.

The Indus Entrepreneur (TiE) was started five years later by the first generation of successful Indian entrepreneurs in the Valley who, like CIE's founders, wanted to make it easier for future generations of Indian emigrants to start businesses. This core group came together in response to a visit from India's Secretary of Electronics to Silicon Valley in 1992. When the minister's flight was delayed, attendees waiting at a conference began to share complaints about the difficulties of running a business in the Valley. The TiE founders chose to call themselves Indus (rather than Indian) entrepreneurs to include other South Asians such as Pakistanis, Bangladeshis, and Nepalese. However, the organization's Bay Area members are almost all Indian. The group included Suhas Patil, former MIT professor and founder of Cirrus Logic; Prabhu Goel, founder of Gateway Design Automation; and Kanwal Rekhi, who started and ran Excelan until it merged with Novell. Like the first generation of Chinese immigrant entrepreneurs, these three had succeeded in spite of their lack of contacts or community support. In the words of another early TiE member, Satish Gupta: "When some of us started our businesses we had nobody we could turn to for help. We literally had to scrounge and do it on our own. What we see in Silicon Valley, especially with the new start-up businesses, is that contacts are everything. All of us have struggled through developing contacts, so our business is to give the new person a little bit of a better start than we had."[10]

Building on Strength

The professional and technical associations organized by Silicon Valley's immigrant engineers in the 1980s and 1990s are now among the most vibrant and active professional associations in the region, with memberships ranging from several hundred in the newer associations to more than a thousand in the established organizations. Annual meetings of the leading Chinese and Indian associations regularly attract up to two thousand attendees and feature some of the industry's most successful executives, entrepreneurs, and venture capitalists along with leading technology and management specialists.

These organizations combine elements of traditional immigrant culture with distinctly high-tech practices: they simultaneously reinforce ethnic identities within the region and facilitate the professional networking and information exchange that help professionals succeed in the highly mobile Silicon Valley economy. They are not traditional political or lobbying organizations; with the exception of the Asian American Manufacturers Association (AAMA), which occasionally has taken public positions and lobbied for Asian-American issues, these groups are oriented almost exclusively toward furthering the professional and technical advancement of their members. Nor are they organizations of elites; most are relatively open organizations with low membership fees that welcome nonmembers (even nonethnics) to their events.

It is noteworthy that the region's Chinese and Indian immigrant engineers organized separately from each other, as well as from Silicon Valley's mainstream professional and technical associations such as the American Electronics Association, the Institute of Electrical and Electronic Engineers, or the Software Entrepreneurs Forum. (It is true that immigrant engineers do join mainstream organizations, but they appear to be less active there than in the ethnic associations.) There is also virtually no overlap in the membership of Indian and Chinese professional associations, although there appears to be considerable overlap within the separate communities, particularly the Chinese community, which has many differently specialized associations with different constituencies. In 2002, Indian and Chinese associations initiated jointly sponsored events in the Valley for the first time.

The Indian and Chinese professional and technical associations are the largest and most visible in the region because these groups dominate the immigrant technology workforce. Associations representing other immigrant engineers are not only significantly smaller but also more recently established, mostly since 2000. By 2004, Silicon Valley was home to more than thirty immigrant professional and technical associations (see Appendix A for a complete list). On the basis of estimated size, the top ten are the following (listed in order of their date of founding):

- Chinese Institute of Engineers (CIE/USA). Founded in 1977, CIE had 1,000 members in Silicon Valley in 2004. It is a technical organization that promotes communication and interchange of information among Chinese engineers and scientists.
- Asian American Manufacturers Association (AAMA). Founded in 1979, AAMA had 1,100 members in Silicon Valley in 2004. It is a multi-technology business network promoting the success of the region's Asian-American technology enterprises.
- Silicon Valley Indian Professionals Association (SIPA). Founded in 1987, SIPA had 2,300 members in Silicon Valley in 2004. It is a forum for young entrepreneurial expatriate Indians to contribute to cooperation between the United States and India in technology areas.
- Chinese Software Professionals Association (CSPA). Founded in 1988, CSPA had 2,000 members in Silicon Valley in 2004. It promotes technology collaboration and facilitates information exchange in the software profession.

- Silicon Valley Chinese Engineers (SCEA). Founded in 1989, SCEA had 4,000+ members in Silicon Valley in 2004. It promotes entrepreneurship and professionalism among members and seeks to establish ties to mainland China.
- Monte Jade Science and Technology Association (MJSTA). Founded in 1990, MJSTA had 470 individual members and 180 corporate members in Silicon Valley in 2004. It promotes business cooperation and long-term investment between Taiwan and the San Francisco Bay Area.
- Chinese American Semiconductor Professionals Association (CASPA). Founded in 1991, CASPA had 3,500 members in Silicon Valley in 2004. It promotes technical communication, information exchange, and collaboration among semiconductor professionals.
- The Indus Entrepreneur (TiE). Founded in 1992, TiE had 2,000 members in Silicon Valley in 2004. It fosters entrepreneurship by providing mentorship and resources.
- Chinese Information and Networking Association (CINA). Founded in 1992, CINA had 2,600 members in Silicon Valley in 2004. It is an organization of Chinese professionals who promote technologies and business opportunities in information industries.
- North American Chinese Semiconductor Association (NACSA). Founded in 1996, NACSA had 1,000+ members in Silicon Valley in 2004. It is devoted to professional advancement in the semiconductor sector and interaction between the United States and mainland China.

All of these associations combine socializing with support for professional and technical advance.[11] They tend to remain independent of one another and of the region's more established groups, in spite of the striking similarity of their missions. Virtually all are committed to promoting entrepreneurship as a means of professional advancement within the immigrant community. Typically, a few successful technology entrepreneurs or executives provide the stature as well as the initial resources for the organization, assuming the role of the "old men" who devote their experience and time to organizing and mentoring subsequent cohorts of immigrants. Not surprisingly, most of these mentors are male. There are some very successful Indian and Chinese immigrant women entrepreneurs in the region today, but they remain exceptional.[12] The oldest associations—the CIE and the AAMA, both founded in the late 1970s—along with a dozen or so groups founded in the 1980s and early 1990s, have seen successive rounds of leadership and successful entrepreneurs. The newer associations remain dominated by their founders.

A typical format for many of the regular association meetings is a presentation by a successful entrepreneur or manager, usually but not exclusively from within the community, and a meal with socializing. For example, a 1994 meeting of the Chinese Software Professionals Association (CSPA) featured Jeff Lin, the general manager of Asante Technology Inc., speaking on the topic "Starting up a Hi-Tech Company and Bringing It Public." Similarly, in 1997 SIPA sponsored a talk by Vivek Ranadive, founder and CEO of TIBCO Inc., titled "The Making of a Networking Software Company." In both cases, admission to hear the speaker was free for members and a minimal amount ($5 or $10) for others, and memberships were equally affordable, around $30 a year. These associations depend heavily on the time of volunteers and contributions from companies.

These professional associations also help to create and reinforce collective identities among immigrants. When Hong Chen, one of the earliest mainland Chinese immigrants in Silicon Valley to start a company and take it public on NASDAQ, spoke in 2000 at a meeting of the Hua Yuan Science and Technology Association (an association of mainland Chinese entrepreneurs), he presented a brief history of four generations of Chinese entrepreneurs in the United States starting in the 1960s and 1970s and elaborated on the growing network of successful Chinese-founded companies. Included in the fourth generation (2000 and later) were three publicly held and twenty-five private companies. Chen also described the growing support infrastructure in the region for Chinese entrepreneurs, including sixteen NASDAQ-listed companies with Chinese CEOs or founders and a long list of promising private companies, venture capital partners, and angel investors with Chinese backgrounds, as well as lawyers, fund managers, and so forth.

The associations sponsor speakers and conferences that provide forums for sharing specialized technical and market information, as well as basic training in the nuts and bolts of entrepreneurship and management for engineers with limited business experience. The Indus Entrepreneurs' second annual conference, TiEcon '97, was titled "Growing an Enterprise Successfully: A 'Mini MBA' for Entrepreneurs"; it included panels on such topics as "Translating an Idea into a Business Plan," "Raising Money for Different Growth Stages," and "Legal Aspects of Starting a Company." Likewise, the theme of the Monte Jade Science and Technology Association's 1999 Annual Conference was "Beyond Cutting-Edge Technologies: What It Takes to Build A Successful Company" and featured presentations on "Strategies for Market Leadership" and "How to Choose an Investment Banker," among others. Both TiEcon and Monte Jade's annual meetings typically attract at least a thousand attendees—and often more.

Forums like this have helped to promote the adoption of Western management models rather than traditional Asian business models based on family ties and obligations (Chinese) or hierarchical authority (Indian) in immigrant-run technology companies. When Min Zhu, co-founder, President, and CTO of WebEx Inc., spoke at the weekly meeting of the Chinese Entrepreneurs Club (which became Hua Yuan in 1999), he cautioned an audience of technically trained immigrants that technology is not all that it takes to start a successful company. His talk, titled "Make the Right Connections in the Right Environment—My Experience as a Technical Person Starting Up Two Companies," focused on the importance of working with experts in sales, marketing, finance, and other nontechnical areas in order to succeed in a start-up. In addition to providing advice on how to start a business, some of the Chinese associations even offer forums on English communication, negotiation skills, and stress management.

Some of Silicon Valley's immigrant associations are technically oriented or have special interest groups organized around particular platforms, technologies, or technical issues. The North American Taiwanese Engineers' Association, for example, had ten specialty technical groups in 1997, including Biotechnology and Science, Computer Networking, Environmental Science, Multimedia, and Photonics Engineering. The association sponsored technical seminars—for example, a 1999 presentation titled "Data Mining on Time Series: Predictive Rules in Economic and Stock Data" by T. Y. Lin, a professor of computer science at San Jose State University. Similarly, the Chinese American Semiconductor Professionals Association (CASPA) and the CIE held a joint seminar titled "IC Packaging Technology and Trends" in the late 1990s, offering four hours of detailed discussion ranging from interconnection issues and high-end packaging for high-performance ICs like microprocessors to the future of microminiaturization.

Looking Outward

These associations try to avoid creating self-contained ethnic enclaves. Chinese and Indian engineers often work closely with immigrants from other countries as well as with native-born engineers. There is growing recognition that while ethnic networks may help in creating a start-up, a new company must quickly become part of the mainstream in order to have access to talent, capital, or partners. In fact, Chinese entrepreneurs frequently make disparaging remarks about family firms. The most successful immigrant businesses in Silicon Valley today are those that draw on ethnic resources, at least initially, while integrating over time into mainstream technology and business networks. This parallels Granovetter's notion of balancing coupling and decoupling in the case of overseas Chinese entrepreneurs.[13]

Some immigrants in Silicon Valley have relied on informal social networks rather than formal organizations, although by now most have at least established a presence on the Internet. Israel's Technion Institute of Technology established a local branch in the early 1980s, but the Israeli technical community in Silicon Valley had no formal association throughout the 1980s and 1990s. Instead, the Israel Economic Consulate in San Francisco and the San Jose Jewish Federation coordinated a loose, informal network of local technology professionals. A subset of this group was dubbed the "California Israel Angels" because they invested in start-ups with a connection to Israel. Informal social networks like these typically build on shared identities created in the home country. The shared experience of military service, for example, frequently serves as a powerful bond among Israeli engineers.

Jean-Louis Gassée, a high-profile senior executive at Apple Computer, was likewise at the center of a network of French engineers and professionals who met and socialized regularly for more than a decade. A self-identified network of Iranian entrepreneurs met informally in Silicon Valley for more than a decade prior to the formation of SiliconIran in 2002. The newly formed group attracted over four hundred participants to its first Iranian-American technology forum.

There is anecdotal evidence that many other informal immigrant networks come and go in Silicon Valley without being institutionalized. Foreign-born graduates of Stanford University, for example, appear to develop particularly strong bonds and also to learn about entrepreneurship together. Victor Wang, who founded GWCom, relates that when he arrived at Stanford University after an exhausting thirteen-hour flight from China, he was invited to a gathering of Chinese electrical engineering students. Rather than an orientation for new students about how to open a bank account or find housing, the meeting turned out to consist of an intense discussion about how to start a company. Wang reports that he was very impressed, and in the coming years he and his engineering classmates developed a strong mission. In his words, "no Chinese could be as well positioned as we were, as Stanford electrical engineering graduates, to start businesses in Silicon Valley."[14]

Jason Wu, founder of Netfront Communications, describes a similar experience:

> Stanford and Xerox PARC had a huge impact on me. At Stanford we had business courses for engineers and leading entrepreneurs visiting our classes, and at PARC there was lots of excitement about pulling a business plan together. I really wanted to build a company. For a while I met with a few friends every Sunday to talk about business ideas. It was a network of Stanford and PARC Chinese engineers. We would write out lots of crazy business plans and talk about how to file patents or how to start a business. We all learned from each other and we learned what it takes to build a company. One guy stayed in his company, another guy started his own company because he wanted to lead, and others just kept hopping jobs.[15]

Alumni Ties

Alumni relationships are among the most powerful sources of collective identity for Silicon Valley's immigrants—many of whom graduated from the same elite engineering programs. This is particularly true in the Chinese community. The National Taiwan University alumni association, with over a thousand members in Silicon Valley, is one of more than a dozen Chinese alumni associations with active branches in the Bay Area. These alumni ties are explicit: the Silicon Valley Chinese Engineers Association (SCEA) Web site, for example, describes its origins in the spring of 1989 "from two groups of early Silicon Valley settlers from China . . . originally from Qinghua University and the Graduate School of the University of Science and Technology in China." The various branches of the Indian Institute of Technology, such as IIT–Bombay, Kanpur, Madras, and New Delhi, also host regular alumni events in the region and provide a strong source of informal networking as well.

Although alumni ties among the graduates of Israel's Technion remain primarily informal, they frequently provide the basis for founding teams of start-ups as well as recruiting and mentoring networks in the region. For some, these alumni ties are more important when in a foreign country than when they are at home. Taiwan-born Fred Cheng, who has lived in Silicon Valley for two decades, reports: "The National Taiwan University (NTU) electrical engineers network is stronger in the United States than in Taiwan because we are all outsiders here. In this industry we tend to rely on our college classmates more than our family ties for help."[16] Jimmy Lee, founder of the semiconductor firm Integrated Silicon Solution Inc. (ISSI), adds: "About forty of my classmates from the 1977 graduating class at NTU are still in the Bay Area. I'm in close touch with about a dozen, but I can call any of them very easily if I need to. For example, two of them are developing network products that require high-performance chips like ours so we regularly discuss the direction of our products."[17]

The Tsinghua Alumni Association of Northern California (which includes both the Beijing and Taiwan branches of the university) has about two thousand members who meet twice a year for large family social get-togethers, as well as on an ad hoc basis to host visiting professors or government VIPs.[18] Tsinghua alumni frequently mentor each other, start companies together, and report that they turn to one another frequently for operational advice or contacts. They also know that they can tap into a pool of talent and connections in China, including students and teachers at their alma mater as well as alumni who are government officials.

In 2001, reporters from the *Electronic Engineering Times* identified a cluster of Silicon Valley start-ups established by Tsinghua graduates in fields including network switching fabrics, software platforms for interactive TV, wireless mobile communications, advanced optical medical equipment, and digital TV modulation chips.[19] Netscreen Technologies, for example, was founded by Beijing Tsinghua graduate Ken Xie and a classmate, and financed by another Tsinghua graduate, to develop high-speed, ASIC-based firewall/virtual private network (VPN) solutions.

A partial list of the Bay Area chapters of foreign university alumni associations suggests the extent of the phenomenon:

- Tsinghua (Taiwan and Beijing together) Alumni Association of Northern California
- Peking (Beijing) University Alumni Association of Northern California (PKUAANC)
- Silicon Valley Tsinghua Network (Beijing)
- Chiao Tung University Alumni Association of Northern California (CTUAA/JTUAA of Northern California)
- Zheda Alumni Network (Zhejiang University Alumni Association in San Francisco)
- The Fudan University Alumni Association of Northern California (FDAANC)
- University of Limerick Alumni Association: Silicon Valley Chapter
- Tongji University Alumni Association of America
- Joint Alumni Association of Chinese Universities and Colleges in Northern California (JAACUC)
- Shanghai Jiao Tong University Alumni Association of Silicon Valley (SJTU-SV)

- Beijing Institute of Technology Silicon Valley Alumni Association
- National Taiwan University Alumni Association of Northern California
- Indian Institute of Technology (IIT) Madras Alumni Association of North America: Northern California
- IIT Bombay Heritage Fund: San Francisco Bay Area Chapter
- San Francisco Bay Area NanKai Alumni Association (NAA-SFBA)
- North American Beijing Normal University Alumni Association
- Jilin University Alumni Association of Silicon Valley
- Nanjing University Alumni Association: Bay Area Chapter

Educating a New Generation

Several alumni of Taiwan's National Chiao Tung University with U.S. graduate degrees have gone beyond alumni associations and started a university to provide training that serves the immediate needs of the local technology community. The founder and first president of Silicon Valley University, Dr. Jerry Shiao, reports: "We saw a big gap between what the local community colleges and other educational institutions were providing and the needs of the Silicon Valley high-tech industry."[20] He and his classmates from Taiwan who were teaching networking and computer science courses in the local community colleges realized that these colleges were not attracting top faculty or providing up-to-date theories and skills; the pay was one-third of local industry salaries or less, so the schools were getting only older employees from established firms, who were teaching with out-of-date textbooks.

In establishing Silicon Valley University, Dr. Shiao and his colleagues recognized that because technology changes so rapidly, it is essential to get instructors directly from the leading companies to ensure that students have skills that are competitive for careers in local high-tech industries. After raising funds from local Chinese entrepreneurs in 1997, they rented space, incorporated the university as a private nonprofit institution, and received state authorization to grant degrees. Dr. Shiao reports that it was easy to find faculty from the alumni networks of National Chiao Tung University. All fourteen of the original faculty members had Ph.D.s from American universities, and twelve were graduates of National Chiao Tung University. They continued to work full time while teaching in the evenings, and they attracted students who wanted to update their skills. The program offers both Master's and Bachelor of Science degrees in Computer Science and Computer Engineering, as well as certificate programs in computer network and telecommunications, database design and software, and English as a second language.

In 1999 Silicon Valley University had fifteen students, all of them Chinese. Although the official language is English, a majority of the students continue to be of Chinese origin—including a growing number who apply from outside the country. The student body has more than tripled in size, and the faculty remains entirely Chinese. Several teachers are now from mainland China, as are many of the students. The university prides itself on being more flexible and adaptable than other educational institutions. It also provides extensive hands-on learning in labs with state-of-the-art equipment (donated from local companies like Cisco, Cadence, Sun, Oracle, and Wind River), including an electronic resource center for digital research, a computer-aided design lab, and a data communication and telecommunications lab. With its small but focused program that improves students' career opportunities while also responding to the changing skill requirements of local high-tech employers, Silicon Valley University is a model of the kind of institution building that is the essence of the Valley's industrial system.

ETHNIC PROFESSIONAL IDENTITIES AND DIASPORAS

The Chinese community in Silicon Valley is distinguished from the broader Chinese diaspora or "overseas Chinese business networks" by shared professional and ethnic identity as well as by deep integration into the Valley's technical networks. These communities differ from the "epistemic" communities of science because they are oriented toward career and entrepreneurial achievement rather than pure scientific progress and are thus open to many types of professionals. They are more akin to "communities of practice" in that they support joint problem solving and learning through practice; however, the social context and shared identities have strong ethnic as well as professional and technical components.

The Chinese Experience

Although the Chinese Institute of Engineers (CIE) was the first organization of Chinese professionals in the Bay Area, the region already had an infrastructure of Chinese associations. San Francisco's Chinatown, the center of Chinese immigration to the area for more than a century, was home to hundreds of traditional Chinese ethnic associations, regional and district hometown associations, kinship groups (clan, family, or multifamily), and dialect associations. Business and trade associations also supported the thousands of traditional ethnic businesses located in the city, including apparel contractors, jewelry and gift shops, neighborhood grocers, Chinese laundries, and restaurants.

CIE distinguished itself both by the social and economic background of its members and by geography. CIE members were highly educated professionals who had immigrated in recent decades from Taiwan or China and who lived and worked in the South Bay area. They had little in common with the older generations of farmers and manual workers who had immigrated from Hong Kong and southern China (Guangdong and Fujian provinces) and who lived and worked in San Francisco. During the 1970s, gatherings of Silicon Valley's Chinese engineers were held in the city. By the mid-1980s, however, as the area's Chinese population increased significantly (and with it the number of Chinese restaurants suitable for meetings!), the center of gravity for socializing shifted decisively to the south. Interviews confirm that these two communities of Chinese immigrants coexist today in the Bay Area, but with limited social or professional interaction.

This divide underscores the dangers of overstating the power of race or nationality in creating cohesive ethnic identities. Collective identities are constructed over time, often through the kinds of face-to-face social interactions that are facilitated by geographic, occupational, or industrial concentration. The initial social connections often have a basis in shared educational experiences, technical backgrounds, language, culture, and history. Once established, these concentrations promote the frequent, intensive interactions that lead to a sense of commonality and identification with members of the same group—and at the same time exclude others, even those of similar racial characteristics.

Ethnic distinctions within the Chinese technology community are also reflected in these associational networks. The Monte Jade Science and Technology Association, for example, originally used only Mandarin at many of its meetings and social events, thus excluding not only non-Chinese members but also overseas Chinese from Hong Kong or Southeast Asia. Likewise, the mainland Chinese who arrived in Silicon Valley about a decade after their Taiwan-born counterparts often set up their own organizations rather than joining existing ones. Mainland Chinese engineers, for example, organized the North American Chinese Semiconductor Professionals Association (NACSA) in 1996 even though the Chinese American Semiconductor Professionals Association (CASPA) had been created five years earlier. The mutual mistrust that characterized relations between mainland and Taiwan-born Chinese in the Valley is not surprising given the different economic, political, and cultural environments from which they come.

Interestingly, this tension seems to be diminishing over time as a result of common economic interests in building the Chinese economy. More senior managers and entrepreneurs share a strong incentive to collaborate on building business ties to the mainland. This shift is reflected in the growing attendance of mainlanders at Taiwan-founded association meetings, and vice versa.

The Indian Pattern

Immigrant groups have not always followed the same path in developing associational patterns. Silicon Valley's Indian immigrants did not mobilize until a decade later than their Chinese counterparts because they were later in achieving a critical mass in the region. When they did organize, they created new associations such as SIPA rather than joining existing groups like the Asian American Manufacturers Association. The reason is doubtless that they felt more comfortable with other Indians, despite the fact that they were often from different regions of the country and spoke different dialects. (In fact, a sizable subset of these engineers grew up in Africa and never lived in India.)

The two Silicon Valley–based Indian professional associations, TiE and SIPA, have continued to grow as the local Indian community expands. These groups create common identities among an otherwise fragmented nationality; Indians are historically deeply divided and typically segregate themselves by regional and linguistic differences (Bengalis, Punjabis, Tamils, Gujaratis, and so on). In Silicon Valley, however, it appears that the Indian identity has become more powerful than these regional distinctions. As V. S. Naipaul wrote of his own upbringing in Trinidad, "In these special circumstances overseas Indians developed something they would have never known in India: a sense of belonging to an Indian community. This feeling of community could override religion and caste."[21] There are of course subgroups with varied amounts of familiarity and trust between them, but the shared experience of immigration appears to make national identities stronger than they would be at home.

The importance of ethnic distinctions can diminish with time and experience, as seen with the assimilation of successive waves of immigrants to the United States. Frequent social or workplace interactions can lead to the redefining of identities and worldviews; for example, the initial distrust and tensions between Taiwan- and mainland-born Chinese in Silicon Valley diminished as the two groups saw growing opportunities to collaborate professionally. Just as the experience of immigration contributed to the creation of a collective identity among the overseas Indian community, the vision of developing the China market proved to be stronger than the earlier ethnic divisions among Silicon Valley's Chinese immigrants.

THE BENEFITS OF LOCAL ETHNIC NETWORKS

Although the economic benefits of immigrant networks cannot be demonstrated definitively, the growth of ethnic professional associations in Silicon Valley has corresponded with a growing visibility and success of Chinese and Indian businesses. Entrepreneurs often give the networks credit for helping first-generation immigrants to overcome their sense of isolation. According to Ronald Chwang of Acer, "Social ties are critical to feeling comfortable in Silicon Valley. All of the technology and people you need to know are here, and the networks help you feel as if you are in front of the technology waves and not being left behind or isolated."[22] In the 1999 Annual Report of the Monte Jade Science and Technology Association (West), the chairman states: "Monte Jade holds as part of its mission the goal of connecting our members, so that together we can sail the waves of change through mutual support and collaboration." Jesse Chen, managing director of the investment firm Maton Ventures, similarly notes:

> The benefit of working with Monte Jade is learning from one another. We often face similar problems and challenges and so we can help one another. When I was a manager I made a mistake and was a loner. It is very helpful to talk about how to look at the market, how to solve problems. Now that I'm a venture capitalist I want to invest in companies and grow CEOs. I won't get involved in daily management decisions, but will help avoid the mistakes that I made and share the learning and experiences of others with them.[23]

These social and professional networks contribute as well to building the confidence needed to assume risk. According to Mohan Trika, CEO of an internal Xerox venture called inXight:

> Organizations like TiE create self-confidence in the community. This confidence is very important . . . it provides a safety net around you, the feeling that you can approach somebody to get some help. It's all about managing risk. Your ability to manage risk is improved by these networks. If there are no role models, confidence builders to look at, then the chances of taking risk are not there. That's what we are saying: "come on with me, I'll help you." This quickly becomes a self-reinforcing process: you create five or ten entrepreneurs and those ten create another ten.

Networking creates real business value as well, he says:

> I can approach literally any big company, or any company in the Bay Area, and find two or three contacts . . . through the TiE network I know so-and-so in Oracle, etc. . . . Because we are a technol-

> ogy selling company for the next generation of user interface, every major software company or any software company must have at least two or three Indians or Chinese employees . . . and because they are there, it is very easy for me, or my technical officer, to create that bond, to pick up the phone and say: Swaminathan, can you help me, can you tell me what's going on . . . he'll say don't quote me but the decision is because of this, this, and this. Based on this you can reformulate your strategy, your pricing, or your offer. . . . Such contacts are critical for start-ups.[24]

A former president of the AAMA describes further advantages of the ethnic networks: "Doing business is about building relationships, it's people betting on people, so you still want to trust the people you're dealing with. A lot of trust is developed through friendship and professional networks like school alumni relations, business associations, and industry ties."[25] David Lam's description is similar: "If there is someone that I know . . . if we have some mutual business interest, then the deal can come together rather fast. And if we have known each other for some years and a certain level of mutual trust has already been established, it is much easier to go forward from there. . . . So I think these connections play a very important role."[26]

The ethnic networks also provide an important source of role models and mentors for newly arrived immigrants. Gerry Liu, who co-founded Knights Technology with four friends from Taiwan, reports: "When I was thinking of starting my own business, I went around to call on a few senior, established Chinese businessmen to seek their advice. I called David Lee. . . . I contacted David Lam and Winston Chen. I called up Ta-lin Hsu. They did not know me, but they took my calls. I went to their offices or their homes; they spent time with me telling me what I should or shouldn't be doing."[27]

Some of the benefits are concrete and quantifiable. TiE's most distinctive contribution is its model of cross-generation investing and mentoring. Because of earlier business successes, TiE's founders are able to provide start-up capital, business and financial advice, and professional contacts to a younger generation of Indian entrepreneurs. These engineers claim that one of the biggest obstacles to their own advancement was the bias of mainstream financial organizations, and in particular, the difficulties faced by nonnative applicants in raising venture capital. Like their Chinese counterparts, they felt like outsiders to the mainstream (primarily white and native) venture capital community.

TiE members often take on the roles of mentors, advisers, board members, and angel investors in new Indian companies. One early recipient of TiE funding, Naren Bakshi, presented the business plan for a company called Vision Software in 1995. Within months, TiE members had raised $1.7 million for Bakshi's company. By 2000 Vision Software had sixty employees and had raised additional funding—in fact Bakshi reports that he was approached by more venture capitalists than he needed. This is in keeping with the TiE founders' vision of supporting "diamonds in the rough" and encouraging them to expand by diversifying their funding and integrating into the mainstream technology community.

TiE members also open their own networks to promising newcomers. Chandra Shekar, founder of Exodus Corporation, reports that the help from TiE members extends beyond providing capital and sitting on the board of directors, to serving as a "trusted friend" or even the "brain behind moving the company where it is today." One of the most important contributions provided by these experienced entrepreneurs and executives is access to "entry points" with potential customers or business alliances. Shekar explains, "The Indian network works well, especially because the larger companies like Sun, Oracle, and HP have a large number of Indians . . . you gain credibility through your association with a TiE member . . . for example if HP wants to do business with you, they see that you are a credible party to do business with. This is very important."[28]

Vinod Khosla, a co-founder of Sun Microsystems and now a partner at the venture capital firm Kleiner, Perkins, summarizes the situation this way: "The ethnic networks clearly play a role here: people regularly talk to each other, they test their ideas, they suggest other people they know, who are likely to be of the same ethnicity. There is more trust because the language and cultural approach are so similar."[29]

Reputation and personal trust remain critical in these communities, particularly in dealing with early-stage start-ups. When successful Indian entrepreneurs like TiE's Patil and Rekhi invest in a company, they provide the legitimacy that allows the entrepreneur to get a hearing from the region's more established venture capital funds. Rekhi, who regularly receives business plans from aspiring Indian entrepreneurs, reports that he will not refer them to investors unless he understands and has complete confidence in their business plan. He recognizes that maintaining his own reputation in the financial community is essential to his ability to continue assisting Indian entrepreneurs.

Satish Gupta of Cirrus Logic similarly notes: "Networks work primarily with trust. . . . Elements of trust are not something that people develop in any kind of formal manner. . . . Trust has to do with the believability of the person, body language, mannerisms, behavior, cultural background . . . all these things become important for building trust. . . . Caste may play a role; financial status may play a role." But he adds that although organizations like TiE are instrumental in creating trust in the community, they also create a set of duties and sanctions: "If you don't fulfill your obligations, you could be an outcast . . . the pressure of, hey, you better not do this because I'm going to see you at the temple or sitting around the same coffee table at the TiE meeting . . . and I know another five guys that you have to work with, so you better not do anything wrong."[30] In the words of Chinese investor Shelby Chen, who considers his community the network of Tsinghua graduates in Silicon Valley: "Trust depends on your reputation for playing fair. If you don't behave, you're an outsider."[31]

Ethnic communities always face the risk of insularity. In the 1990s, some suggested that TiE's network remained so closed that it prevented outsiders from participating. According to a charter member of TiE, there is little desire in the organization to turn outward: "This network just does not connect to the mainstream. If you look at the social gatherings that the TiE members go to, it's all Indians. There's nothing wrong with it . . . but I think if you don't integrate as much, you don't leverage the benefit that much." The persistence of negative attitudes toward immigrants in the United States could potentially reinforce this insularity. A 2001 survey, for example, found negative and stereotypical attitudes toward Asian Americans among about 25 percent of respondents.[32] The challenge for Silicon Valley's immigrant entrepreneurs is a modern variation on a familiar theme: to continually balance reliance on ethnic networks with integration into the mainstream technology community.

SILICON VALLEY'S NEW IMMIGRANT ENTREPRENEURS

The popular image of the entrepreneur as an individual loner, an isolated pioneer, is far from the reality of Silicon Valley. Starting a new technology company requires access to a diverse range of human, financial, technical, and professional resources as well as the confidence to take professional and economic risks. As we have seen, Silicon Valley's immigrant engineers, like their mainstream counterparts, rely heavily on informal as well as formal networks that help them mobilize information and know-how, raise capital, recruit talent, and gain feedback.

Ironically, many of the distinctive features of the Silicon Valley business model were created during the 1960s and 1970s by engineers who saw themselves as outsiders to the mainstream business establishment on the east coast. The region's original industry associations—the American Electronics Association, for example—were an attempt to create a presence in a corporate world from which Silicon Valley's emerging producers had been excluded. These organizations provided role models and support for entrepreneurship similar to that now being provided within immigrant communities. The decision to start a company is heavily influenced by relationships with role models, mentors, friends, and colleagues, and the first high-profile successes within any community frequently provide role models for future entrepreneurs from the same background. A process that began with the original Silicon Valley outsiders continues now with groups that likewise regard themselves as outside the mainstream.

First-generation immigrants to Silicon Valley not only socialize but also start companies with others from the same background. Indian and Chinese immigrants, for example, are significantly more likely to start companies with other Indian and Chinese immigrants, typically classmates or former colleagues, than with others. Half the foreign-born entrepreneurs surveyed in Silicon Valley in 2001 reported that their company had two or more founders from the same country of birth.[33] This reflects the obvious and understandable importance of shared language, history, and education in creating ethnic professional identities among first-generation immigrants: starting a company requires a high degree of trust in your co-founders, after all. But while immigrants rely heavily on friends and colleagues from their home countries to start companies, it is also important to emphasize that this ethnic dominance decreases steadily as companies grow. About one-third of companies with fewer than a hundred employees surveyed in 2001 reported a majority of employees from the founders' country of birth, but this share dropped to less than 10 percent in companies with a thousand or more employees.[34] (For more details of the survey, see Appendix B.)

Silicon Valley's skilled immigrants have been quick to catch the entrepreneurial bug. Unfortunately, it is difficult to get accurate estimates of ethnic or immigrant entrepreneurship in technology industries. The standard way to measure immigrant entrepreneurship is to examine the "self-employed" category in the U.S. Census.[35] Although this may provide a good approximation for owner-run businesses in traditional industries, it almost certainly leads to a significant undercount in technology sectors because so many companies are funded with angel or venture capital—and hence are not fully owned by the founding entrepreneur. Identifying businesses with CEOs having Chinese, Indian, or Korean surnames in a Dunn & Bradstreet database of technology firms started since 1980 provides a more accurate estimate of ethnic entrepreneurship in Silicon Valley.[36] These companies collectively accounted for over $25 billion in sales and close to 100,000 jobs.

Like their native-born counterparts, most highly skilled immigrants aspire to become entrepreneurs, and Silicon Valley's open labor markets allow them to gain a range of managerial and entrepreneurial experiences before striking out on their own. By 2001 over half of Silicon Valley's foreign-born professionals reported that they had worked in a start-up company, either part time or full time, and 62 percent reported that they planned to start a business on a full-time basis—compared to only 46 percent of U.S.-born professionals.[37] Nonetheless, there is little difference in the way that skilled immigrants and their U.S.-born counterparts start technology companies. The majority of both foreign and U.S.-born entrepreneurs incorporate their firms in the United States; almost all raise money initially from personal savings and angel investors, and in subsequent rounds from venture capital firms; and they raise comparable amounts of capital. Both groups rank "access to investors" as the most significant barrier they face in raising capital for their start-ups. Immigrant-run firms go public at the same rate as those started by U.S.-born entrepreneurs.[38]

The increased visibility of successful Chinese and Indian entrepreneurs and executives in Silicon Valley during the 1990s transformed their image in the mainstream community as well. Most Asians in the region today say that although the glass ceiling may exist in traditional industries or in larger, old-line technology companies, it has diminished dramatically as a problem in Silicon Valley. In fact, by the late 1990s the region's leading mainstream venture capital firms were clamoring to hire Asian-American partners for fear of losing potentially lucrative deals. Skilled immigrants who experience limits to their professional advancement now have many alternatives, including starting their own companies or returning to their home countries.

In short, by the turn of the century skilled immigrants in Silicon Valley were starting technology companies at the same rate as their native-born counterparts. Foreign-born engineers and scientists learned the SV model of entrepreneurship quickly, and successfully combined their technological capabilities with the venture-financed, high-growth business model that distinguishes the U.S. technology sector. Ethnic professional associations and social networks in turn allowed them to mobilize resources collectively, while simultaneously creating shared identities as engineers-turned-entrepreneurs and as part of the broader technical community.

3 ❖ Creating Cross-Regional Communities

Two decades after it was founded to serve the needs of emerging technology start-ups, Silicon Valley Bank sponsored three high-profile international missions to India, Israel, and China. The bank took a delegation of two dozen Silicon Valley venture capitalists to Bangalore and Mumbai in 2003 for a week of meetings with local investors, entrepreneurs, and business and government officials. Six months later the bank sponsored comparable educational and networking visits for groups of venture capitalists to Tel Aviv, Shanghai, and Beijing. At the same time, the bank announced plans to open international offices in these locations.

Silicon Valley Bank sees its role as helping to create in other regions a local "economic ecosystem" like that in Silicon Valley. Although the bank had informally cultivated contacts in India, China, and Israel for years, these missions represented the first step toward formalizing international relationships and anticipated the bank's decision to offer financial services in those regions. The bank's Web site made its intentions clear:

> One of the primary things your business needs as it grows globally is a network of contacts—venture capitalists and service providers—who can help you make the connections you need to ease your

> global expansion. We know people in the major technology centers around the globe, from Silicon Valley to Shanghai, Tokyo to Tel Aviv, Boston to Bangalore. We understand the unique challenges you face and have the support, expertise, and connections you need to succeed.[1]

The Silicon Valley venture capital community, on the other hand, had shown only modest prior interest in Asia. The exaggerated but telling motto of the Sand Hill Road community during the 1990s was that they would invest "no further than twenty miles from their offices," and to this day, many funds invest most of their capital within the San Francisco Bay Area. A handful of venture capital firms started by Chinese, Indian, and Israeli immigrants focused on international markets, and immigrant entrepreneurs demonstrated growing success in the Valley, but a provincial mindset dominated even those mainstream firms with partners of Chinese or Indian origin.

The expansion of Silicon Valley venture capitalists and banks into Asia institutionalized the informal networking pioneered by the region's immigrants in earlier years. Silicon Valley's foreign-born entrepreneurs, managers, and investors began to establish professional and business connections not only with local peers, but with their home country counterparts as well. In so doing, they transformed the one-way brain drain into a complex, two-way process of "brain circulation" and, over time, transferred the central elements of the Silicon Valley entrepreneurial system to distant regions.

These communities of cross-regional engineers and entrepreneurs, with the support of their home country counterparts, nurtured the early investments that, over time, transformed the business environments of their home countries. Immigrant associations and professional service providers frequently catalyzed the flows of people, information, and know-how that laid the foundations for local entrepreneurship and cross-regional collaboration. By the early 2000s, networks of foreign-born, U.S.-educated engineer-entrepreneurs, investors, and managers had dramatically increased flows of information and know-how as well as technology investment into select regions of these formerly isolated economies, while maintaining close ties with their counterparts in the United States.

In this chapter I give an overview of the spread of the Silicon Valley system and then look at the Israeli experience in some detail; the subsequent chapters focus on the cases of Taiwan, China, and India. These cases are not alone, of course. For example, a similar process occurred in Ireland (on a substantially smaller scale) during the 1990s, when the reversal of the brain drain among technical professionals and the development of a venture capital industry were particularly striking.[2] And indications of a similar process are appearing in Russia, which boasts the world's largest remaining pool of relatively untapped technical skill.

The four cases that I examine in this book provide strong examples of the role of the new Argonauts in creating regional advantage. In Israel and Taiwan, which are relatively small economies, the Argonaut advantage is already clear, while in China and India (which are the source of Silicon Valley's largest skilled immigrant populations) the impact of the new Argonauts will likely be the most far-reaching.

REVERSING THE FLOW

Israeli and Taiwanese engineers began returning home from the United States in the 1980s in response to the receptive environments and professional opportunities in their small, relatively open economies. Within a decade they had transferred the institutional and market knowledge that supported the emergence of dynamic centers of entrepreneurial experimentation and innovation in the Tel Aviv and Hsinchu regions.

Although they represented substantially larger communities, few Indian and Chinese engineers returned home from the United States in this period, deterred largely by the lack of professional opportunities and the lower standards of living in their native countries. The proportion of U.S.-educated science and engineering doctoral students from China and India who intended to return home after graduation continued to fall in the 1990s. By 1998–2001, only 3.8 percent of Chinese and 6 percent of Indians planned to return, compared to 31.2 percent of Taiwan-born graduates and 36.7 percent of Israelis. In the earlier period 1990–1993, 6.5 percent of Chinese and 14.4 percent of Indians planned to return home.[3]

The pressure caused by unprecedented shortages of technical talent in the United States during the technology boom of the late 1990s helped overcome resistance to returning and doing business in India and China, as did the market liberalization and growth of both economies. The challenge of recruiting in Silicon Valley, where salaries for engineers were among the highest in the world, triggered a search among local managers for additional sources of skill. Since the 1970s, local firms had established product design and development, lower-level (second- or third-generation) R&D, and IT and customer support infrastructures across a range of locations both inside and outside the United States. These investments were often in locations with an English-speaking population, such as Israel, Ireland, and Singapore, or in other advanced economies including Canada, Germany, Great Britain, and Scotland. While these places were relatively accessible, they offered limited supplies of labor and cost savings.

Silicon Valley's Indian and Chinese entrepreneurs and managers recognized that there were large pools of underutilized technical skill in their home countries, and they began to take the initiative, and the risk, required to do business in countries with limited prior engagement with the world economy. Chinese universities in the late 1990s were granting 328,124 science and engineering (S&E) first degrees annually, almost equivalent to the 384,674 granted in the United States; and Indian universities awarded 176,036 S&E degrees per year in this period, making them the world's fifth largest producer of S&E undergraduate degrees, after the United States, Japan, China, and Russia.[4]

The main constraint in this environment was not technical talent or capital but language skills, institutional and cultural know-how, and the contacts required to succeed in these foreign economies. Although business and professional connections between Silicon Valley and India and China remained limited during the 1990s, cross-regional traffic proliferated in the following decade. The 2001 stock market correction and the worst technology recession in the history of Silicon Valley pushed hundreds of immigrant engineers to return home in the face of unemployment. Equally important, strong growth rates in China and India made these countries attractive destinations and offered professional opportunities that were often superior to those in the United States.

These connections were not entirely new. During the 1990s, Silicon Valley technology firms had increasingly set up offshore research and development centers in India and China in order to gain access to scarce skills. According to the Chief Technology Officer at Synopsys, a Silicon Valley–based electronic design automation (EDA) firm with R&D centers in Bangalore and Hyderabad as well as in Taiwan, Beijing, and Shanghai, "Most of these sites were established in the late 1990s when we were desperate for talent and would go wherever it took to find that talent."[5] One indication of this trend is that the R&D performed by majority-owned overseas affiliates of U.S. companies (figured in current U.S. dollars) in China grew from $7 million in 1994 to $506 million in 2000, and in Israel from $96 million to $527 million.[6]

Initiating Cross-Regional Connections

The immigrant professional and technical associations described in Chapter 2 paved the way for globalization of the Silicon Valley system. Although in the 1980s these associations focused exclusively on supporting entrepreneurial and professional advancement within the Valley, a decade later their ambitions were expanding to include building connections to their home countries. Alumni associations played a similar role by deepening the social connections that support cross-regional collaboration. With time, the new environment became self-perpetuating—as the home countries became more attractive as workplaces, they attracted more and more interest among potential returnees, including service providers as well as entrepreneurs.

A group of the Valley's most successful overseas Chinese executives established the first explicitly cross-regional association, the Monte Jade Science and Technology Association, in 1989 "to provide a bridge for the flow and exchange of information on technology, management, and investment opportunities across both sides of the Pacific."[7] The growth of cross-regional entrepreneurship can be traced in part to the annual Monte Jade investment conference, started in 1998 to bring together Chinese entrepreneurs and investors from Asia and the United States. A committee of the board of directors also provides assistance with new firm formation and growth; one executive reports building connections with individuals in the Taiwan Stock Exchange to help Silicon Valley companies access the public capital market. By 2000 the group's members included most of Taiwan's leading technology companies as well as virtually all of Silicon Valley's Taiwanese companies.

The founders of the Indus Entrepreneur (TiE) were wary of connections to India in their first half-decade because many had been frustrated in their attempts to do business there. This situation changed suddenly in the late 1990s when business leaders recognized a changing environment for entrepreneurship in India. Co-founder Kanwal Rekhi made more than a dozen trips per year in 1999 and 2000, establishing the first non-U.S. TiE branches in Bangalore, Hyderabad, and Mumbai. The organization's subsequent spread was remarkable: by 2003 there were chapters in ten cities in India and two in Pakistan. A year later, in 2004, TiE had thirty-eight chapters worldwide, including a dozen in India.[8]

Mainland Chinese entrepreneurs in Silicon Valley attained visible successes about a decade after their Taiwan-born counterparts. Many mainlanders knew one another through the alumni networks of technical universities such as Tsinghua and Beijing University, and they started an informal Entrepreneur's Club, or E-Club, in 1997. Two years later the E-Club was renamed the Hua Yuan Science and Technology Association, and its mission was expanded to include building cross-border business and technology relationships with China. (Both the mission and the name appear to be modeled after Taiwan's Monte Jade Science and Technology Association.) The new orientation reflected growing confidence after several mainland Chinese start-ups were listed on NASDAQ or acquired, including GRIC and Internet Image. Hua Yuan immediately established an office in Beijing to build working relationships with Chinese regulatory agencies and business executives and to provide administrative assistance for returnees. The 2002 Hua Yuan Annual Conference, titled "Opportunities and Challenges: Riding the China Wave," attracted more than a thousand people.

After 2000 many other immigrant associations expanded their attention beyond local entrepreneurship and began to build long-distance collaborations with their home countries. These include groups as diverse as the Korean Information Technology Network, the American Association of Russian Expatriates (AmBAR), and Silicon Armenia (ArmenTech). Silicon Valley is also home to associations of immigrants who have little interest in returning to their home countries (often these are political refugees) but still aim to build ethnic technology networks. SiliconIran, for example, seeks to become a center for information exchange among Iranian high-tech communities worldwide. The Vietnamese Silicon Valley Network (VSVN) in 2000 described its mission as becoming the "ambassador of entrepreneurial company-building initiatives for the global Vietnamese high-tech business community." However, past experience suggests that groups like the Vietnamese remain ready to return home under the right circumstances.

The Lure of Home

By 2002, China and India appeared much more attractive than before to native-born engineers working in the United States. This enthusiasm is captured by the words of Myles Lu Chunwei, who left a job at Microsoft to start a business in Beijing: "It's a big trend now, people just want to go back to China. It's like the Gold Rush. They're successful in the U.S. but in their hearts they still feel like immigrants. They feel welcome here in China . . . there are not many new opportunities left in the U.S."[9] The 2002 NACSA "China Meets Silicon Valley High-Tech Expo" attracted some five thousand people to a two-day exhibition, exchange, and recruiting session that featured representatives of twenty high-tech parks and companies along with government officials from all over China.

The brain circulation between Silicon Valley and India initially grew more slowly, perhaps because of the greater distance from California (a 24-hour flight to Bangalore compared to a 12-hour flight to Beijing) or the challenges of India's physical infrastructure. However, a basis for increased circulation was forming in earlier years. For example, the online magazine *siliconindia* was started in the late 1990s with an editorial board composed of high-profile Indian entrepreneurs and executives from both the United States and India;

it rapidly amassed a readership of tens of thousands.[10] *Siliconindia* publishes articles on technology and companies in India and the United States as well as related policies, institutions, and services from immigration law to venture capital to recruiting. The annual *siliconindia* career factory in Silicon Valley brings together managers and recruiters from the leading Indian technology businesses with Indians who are considering or planning to return to work in India.

No more than a trickle of U.S.-educated engineers returned to India in the 1990s. Attitudes changed dramatically, however, as a result of both the technology bust and a growing perception of opportunity in India. The younger, more recent immigrants, who faced tight labor markets and the high cost of living in Silicon Valley, were most likely to return to India, but before long even those of an older generation began to return. One Indian who had worked for Intel in the United States for sixteen years and returned to India in 2004 reports that "the decision [to come back] was based on the opportunity . . . to be part of the globalization excitement in India."[11]

Vinod Gopinath, a former IBM employee, accepted a job at Bangalore-based Insilica, a semiconductor start-up co-founded by Vinod Dham (a very successful Indian engineer who was on the Intel team that designed the Pentium chip): "When you are looking to ride a wave, there can be none better than the one at home. Your options are better. There is more pride. The challenge is greater. You spent all your time [in another country] just proving yourself. Now here was an opportunity to prove not just yourself but your country to the rest of the world."[12]

The Role of Service Providers

As the case of Silicon Valley Bank illustrates, professional service providers were early and active participants in the cross-regional communities. Over time, they have helped to build local institutions for entrepreneurship in their home countries. The region's immigrant lawyers, venture capitalists, accountants, headhunters, and consultants are frequent attendees and speakers at the meetings of the ethnic professional associations, and the service firms are among the largest financial supporters of these associations.

Silicon Valley Bank began to strengthen its relationships in India in the late 1990s by establishing a partnership with TiE and hiring an Indian-American as director of the U.S.-Indus Technology and Venture Capital Network, whose focus is building alliances in India and Southeast Asia. As the new director explains, "We must build these international alliances because our customers are becoming global." John Dean, the former chairman, president, and CEO of the bank, said in a 2000 interview: "Even seven years ago when I joined the bank, we . . . had no contacts or relationships established with clients internationally—we were primarily a northern California-based business." However, he went on, today "entrepreneurial technology companies exist throughout the world and we need to learn about the shifts and changes that are taking place, so we can participate and benefit from it. This is not merely in terms of opening offices in newer markets, but also looking for strategic partners and creating linkages with technology entrepreneurs in those areas because we increasingly believe that the flow in terms of capital, technology, manufacturing, distribution, packaging, and licensing is a worldwide phenomenon. This will enable us to do a better job to introduce our clients to the world, and provide assistance to non-U.S. companies who want to set up business here."[13] Dean also noted how quickly the bank's technology focus has changed from an initial focus on semiconductors and computers to include biotech, software, and later the Internet.

Harry Kellogg Jr., the bank's vice-chairman, underscores the importance of the social networks that grow in part out of participation in local associational activities: "A lot of our partnerships are informal—we don't have any written agreements or memos of understanding—but are based on the years of individual relationships we have developed over the years with investors or financial institutions in the Valley and around the world, venture firms, law firms, accounting firms, and others. So when an entrepreneur comes to us and says, for instance, 'Introduce me to the angel community' or 'Which law firm can help me?' we will be able to help through our relationships."[14] Daniel Quon, senior vice president of the Pacific Rim Group for the Silicon Valley Bank, explains that he depends extensively on these networks: "In all of the [more than ten] years that I've worked in Silicon Valley, I've never had to make a single cold call. Everything I do is through the networks, especially the Chinese network."[15]

Law firms with expertise in technology business issues such as intellectual property rights and the initial public offering (IPO) process have been quick to set up international connections, following their clients and emerging opportunities. Carmen Chang, a Taiwan-born, U.S.-educated partner at Wilson Sonsini Goodrich & Rosati, one of Silicon Valley's leading law firms, began serving the region's Taiwanese entrepreneurs and small firms in the 1990s. She got her first deals because some of these engineers had known her father in Taiwan, and she spoke Mandarin. After a few successful transactions, Chang attracted a fast-growing flow of clients, first Taiwanese companies seeking to go public or invest in the U.S. and in China, and increasingly also mainland Chinese entrepreneurs or firms wanting to enter the U.S. or build companies in China, as well as U.S. investment banks, equity funds, and venture funds seeking to invest in China. Chang reports that once accepted into and part of the local Chinese network, she soon had far too much business to take on by herself.[16]

Lawyers like Chang, as well as other professionals who participate in the cross-regional networks, are also frequently drawn into serving clients and advising on policy and regulatory activities overseas.

They often fly so frequently between distant economies that they spend much of their time in the air—what the Chinese call the "astronaut" life. Chang was recruited in 2001 to lead the legal team for the SMIC foundry investment in Shanghai, a more than $2 billion deal that involved several major U.S., Taiwanese, and Chinese investors. She also helped establish a joint venture between Silicon Valley–based 3Com and China's Huawei, and she served as an adviser to the Chinese Securities and Regulatory Commission as they planned for a NASDAQ-type market in Shenzhen.

Venture Capital Goes Global

A new generation of immigrant venture investors is actively seeding cross-regional collaborations. Like their mainstream Silicon Valley counterparts, many of these immigrant venture capitalists have engineering graduate degrees and experience in technology industries. However, unlike the older generation of venture capitalists, whose networks and investments tend to be very close to home, the immigrant investors see their role as bridging geographically distant centers of skill and technology. In the words of overseas Chinese financier Peter Liu, who has worked in Silicon Valley for more than two decades and now runs the venture firm WI Harper:

> WI Harper distinguishes itself from its Sand Hill counterparts through its personal and professional ties with key management in Asia. . . . In Asia it is very difficult to get good information, and through our established network of contacts we are in an excellent position to help the companies in which we invest. . . . We see ourselves as the bridge between Silicon Valley and Asia.[17]

This requires a depth of understanding of not only the Silicon Valley business culture, community, and institutions, but also those of the multiple—and highly differentiated—nations of Asia. Institutions, languages, and culture vary significantly between the countries of the region, and in large countries like India and China there are important regional differences as well. The regions often have distinct dialects and even languages as well as differing cost structures, markets, infrastructures, regulations, and so forth, all of which require localization and, in some cases, different business models.

According to Herbert Chang, partner at the Taiwan-based venture firm VentureStar, who in the 1990s traveled across the Pacific ten or twelve times per year: "We are matchmakers between Silicon Valley and Taiwan. We provide market information for both. We provide intelligence for Taiwanese companies because we screen thousands of U.S. start-ups every year. They know we have a good track record and reputation, know that we'll bring them good stuff. We also provide contacts and information for Silicon Valley companies because I know their potential customers and vendors so well that we don't need financial statements and a long due diligence process before setting up partnerships."[18] This matchmaking role works best, of course, in the context of ongoing and trust-based relationships in which the collaborators have shared backgrounds and worldviews of the sort that characterize Silicon Valley's immigrant technical communities.

The great majority of the immigrant-run venture capital firms remain focused on their own home country markets because of the need for local connections, linguistic fluency, and institutional knowledge. However, a few have succeeded in becoming international, primarily through skillful selection of capable multicultural professionals and high-caliber partners in the relevant countries, as well as the ability to work with minimal hierarchy and allow substantial autonomy to national offices. Walden International Investment Group (WIIG), for example, was founded in 1987 by Lip-Bu Tan, a Singapore-born Chinese American. Through early investments in Taiwan and Singapore, Walden helped to introduce and pioneer the

U.S. venture capital concept in Asia. In 2005 Walden had committed capital of over $1.6 billion and had offices in Taiwan, China, Singapore, Malaysia, and India as well as the San Francisco Bay Area. About a dozen other Silicon Valley firms invest primarily or exclusively in Asian technology ventures.[19] These firms invest alongside and with networks of angel investors, corporate venture capital funds (such as Intel Capital), domestic venture capital firms, and later-stage investors such as Goldman Sachs (Asia).

As lawyers, venture capitalists, investment bankers, entrepreneurs, managers, and other professionals travel between regions, they transfer technical and institutional knowledge as well as contacts, capital, and information about business opportunities and markets. Information moves rapidly between distant regions because of the density of the social networks, shared identities, and trust within the communities. In the words of a Taiwanese engineer in 1996, "If you live in the United States it's hard to learn what is happening in Taiwan, and if you live in Taiwan it's hard to learn what is going on in the U.S. Now that people are going back and forth between Silicon Valley and Hsinchu so frequently, you can learn about new companies and new opportunities in both places almost instantaneously."[20]

Information about the success (or failure) of new ventures travels particularly quickly through the immigrant social networks. Successes—when a company grows rapidly, goes public, or is acquired—serve as role models and encourage subsequent experiments with return migration or investment. These events can become turning points under the right circumstances, as with the accelerated return of Taiwanese and Israelis to their home countries in the 1990s and the returnees to China and India a decade later.

In addition to permanent returnees, a growing number of immigrant professionals work in *both* places, traveling between Silicon Valley and their home country once or twice a month. While this is only possible because of the improvements in transportation and communications technologies, it does not mean that these perpetual travelers are rootless. Their personal networks and knowledge of local institutions in both Silicon Valley and their home countries are essential to initiating collaborations between individuals and firms in the two regions.

CROSS-BORDER COLLABORATIONS AND REGIONAL ECONOMIES

Returning entrepreneurs and their communities transfer elements of the Silicon Valley business model to their home countries, including venture capital finance, niche market specialization, professional management, minimal hierarchy, merit-based competition, and reliance on external partners. This almost always represents a significant departure from the established business models in these economies, which are typically either large, diversified family businesses or state-owned enterprises. Returnee entrepreneurs and investors explicitly avoid family-run or state-controlled business as they create local models of entrepreneurship.

The growing linkages between Silicon Valley and emerging technology regions not only stimulate new local firm formation, but also generate increasing opportunities for long-distance collaborations. This process fueled entrepreneurship-led growth in both Taiwan and Israel during the 1980s and 1990s, and it is accelerating in select regions of India and China today.

Data on private equity investments and fundraising are indicative of this trend. Israel's venture capital industry is typically regarded as second only to its U.S. counterpart, despite its small population. From 1999 to 2002, Israel ranked first in the world in venture capital investment as a percentage of GDP, well ahead of the second- and third-ranked United States and Canada.[21] Taiwan, which also boasted over $1 billion (in U.S. dollars) in venture capital investments in the late 1990s, was ranked first in Asia in private equity investments. By 2003 the private equity invested in China had reached $1.3 billion, while in India the amount reached $0.8 billion.[22] The total market capitalization of the Taiwan Stock Exchange exceeded $350 billion in 2004, making it the world's largest emerging stock market; the 2004 figures for the stock markets in India and China were approximately $250 billion and $190 billion respectively, and they were growing rapidly.[23]

Not for the Weak

These emerging regions are populated by more than start-ups and small- and medium-sized firms, to be sure. By 2000 Israel and Taiwan were home to thousands of technology enterprises, all started since 1980. The most successful start-ups grew quickly and, like Hewlett-Packard or Fairchild in Silicon Valley's first decades, served as crucial sources of managerial and technical training as well as entrepreneurial spin-offs. Taiwan's Acer, Israel's Checkpoint Software, and India's Wipro have played this role in their local economies. The departure of senior managers or engineers to join local start-ups or start new enterprises contributes to local learning as well as to the diversification of the regional economy.

These regions are home to U.S. and other foreign technology companies as well—typically the R&D labs and corporate venture capital arms as well as manufacturing or development activities. Their relationship to the local economy differs from that of the vertically integrated modern corporation of an earlier era, with its internal labor markets and commitment to corporate secrecy and self-sufficiency. This is not an issue of firm size or scale but rather of organization and strategy. The most successful competitors in technology markets today recognize that innovation can happen almost anywhere. These firms do not grow by controlling sources of supply or markets, but rather through specializing and engaging in shifting collaborations, either local or long distance, with other specialized producers.

Large corporations, both domestic and foreign, also play a central role as venture investors in cross-border enterprises. Intel Capital, Acer Technology Ventures, Motorola, and TSMC all invest aggressively in start-ups as much for access to potentially strategic new technologies as for financial return. However, investing wisely in distant regions requires up-to-date local knowledge and hence integration into local as well as cross-regional networks.

Integrated Silicon Solution Inc. (ISSI) was one of Silicon Valley's earliest cross-border start-ups. Taiwan-born Jimmy Lee earned a graduate degree in electrical engineering at Texas Tech in the late 1970s and worked in the Silicon Valley semiconductor industry for over a decade before joining K. Y. Han, a National Taiwan University classmate, to start ISSI. They bootstrapped the start-up with their own funds and those of Taiwan-born colleagues, and subsequently raised over $9 million from venture capital funds managed by overseas Chinese, including Walden International and H&Q Asia Pacific.

Lee and Han exploited resources in both Silicon Valley and Taiwan to grow ISSI. They recruited former colleagues and classmates to the R&D center in Santa Clara to design high-performance CMOS memory chips, and they lined up a manufacturing partnership with the recently established Taiwan Semiconductor Manufacturing Corporation (TSMC). They also incorporated a subsidiary in Taiwan's Hsinchu Science Park to oversee assembly, packaging, and testing. Han traveled to Taiwan monthly to monitor the firm's manufacturing operations, but he soon decided that he needed a greater presence in Taiwan and moved his family home to run ISSI–Taiwan. Lee remained in Silicon Valley as CEO and chairman.

ISSI grew rapidly in the early 1990s, selling its high-speed Static Random Access Memory chips to motherboard firms in Taiwan. Not only did Lee and Han have the language skills and cultural know-how to sell in this market, but they benefited from the fact that many of their classmates from National Taiwan University had moved into leadership positions in local industry. In 1995 ISSI was listed on NASDAQ, making it one of the first Silicon Valley companies started by overseas Chinese to go public in the United States. Several years later Lee and Han spun off ISSI–Taiwan as an independent venture, Integrated Circuit Solution Inc. (ICSI), so that it could go public on the Taiwan Stock Exchange.

This pattern of specialization and decentralization is characteristic of the loosely coupled networks of firms that are emerging in today's global economy. ISSI still owns a minority share of ICSI, and the firms do not compete: ICSI specializes in designing semiconductors for the PC and peripherals markets, while ISSI has shifted into very high-performance ICs for networking, telecommunications, Internet, wireless, and handheld applications that demand portability, connectivity, and increased bandwidth. ISSI has also invested in other promising enterprises in Silicon Valley and Taiwan: the firm owns 14 percent of NexFlash Technology Inc., a U.S. flash memory firm; 17 percent of GetSilicon Inc., a provider of collaborative supply chain management solutions; and 26 percent of E-CMOS Corporation, a fabless semiconductor company (that is, a company that outsources its chip production) located in Taiwan.

ISSI remains small, with 260 employees and sales of $100 million, yet it is quickly building ties to mainland China. In 2001 the firm invested $40 million in SMIC, a new Shanghai-based semiconductor foundry, to ensure low-cost advanced-process wafer capacity and access to the China market. The original investors in ISSI—H&Q Asia Pacific and Walden International—are also among the lead investors in SMIC. The firm recently established a Shanghai subsidiary located near SMIC with facilities for design, logistics, and sales/marketing. ISSI is clearly not competing on the basis of low-cost labor alone. The firm has its own intellectual property (over seventy U.S. patents) and is highly productive (revenue of $346,000 per employee in 2003).

Cross-regional entrepreneurs like Jimmy Lee and K. Y. Han do not abandon Silicon Valley when they return home; rather they continue to draw upon its distinctive capabilities, while also accessing the skills and resources in regions elsewhere (in their case in Taiwan and, more recently, in China). ISSI keeps a base in California in order to tap its world-class design skill and architectural knowledge as well as to retain access to leading-edge markets for its chips (more than half of ISSI's customers are based in the United States). However, they also rely on Taiwan's sophisticated manufacturing capabilities and China's lower-cost engineering skill. By 2003 ISSI had 170 employees in Silicon Valley, 60 in China, and only 30 in Taiwan.

This is not a task for the weak. Returning entrepreneurs regularly face challenges ranging from opaque bureaucratic regulations and undeveloped infrastructure to corrupt officials. The efforts of the first Chinese firms seeking to raise capital from U.S. public markets in the late 1990s were stalled for months by Chinese regulators who insisted on enforcing a myriad of detailed but irrelevant reporting procedures. And domestic engineers are often resentful of the higher status and economic position accorded to returning engineers. Cultural differences cause problems as well. K. Y. Han describes setting up ISSI–Taiwan in the early 1990s:

> The differences between Taiwan's culture and U.S. culture can create very big problems. In the semiconductor business we need to move very fast, with six- to nine-month product cycles, so we can't afford to have communication misunderstandings. I have a great advantage because I understand the thinking and philosophy of the

U.S. employees because I lived and worked there for a long time, and I also understand the Taiwanese employees because I grew up here. The U.S. employees are far more aggressive and straightforward, while the Chinese employees are more conservative and passive. So if a U.S. guy discusses an engineering problem, the Taiwan employee is likely to take it personally.[24]

Communication problems arise even when employees speak the same language. Taiwanese managers who are setting up factories in mainland China learn quickly that although they speak Mandarin, they still encounter frequent misunderstandings with their mainland-born counterparts. Similarly, conflict within teams is frequent between Indian engineers accustomed to a hierarchical management style and their U.S.-trained counterparts who work in a more confrontational manner. These conflicts are more severe when, as is often the case, collaboration occurs over long distances.

Firms like ISSI often compensate for institutional problems in their home countries by incorporating in the United States to take advantage of the stability and maturity of American legal and financial institutions, and maintaining headquarters or R&D offices in Silicon Valley even when manufacturing and development take place in their home country. Many continue to recruit managerial and technical talent from the United States and invest in start-ups or build partnerships and alliances with firms in Silicon Valley to stay abreast of technology and market developments. Some, like ISSI, also take advantage of U.S. capital markets and go public on NASDAQ.

Even though they mimic its management and entrepreneurial models, the technology regions in Taiwan and Israel are not becoming replicas of Silicon Valley. Rather, these economies have coevolved through diverse collaborations with differently specialized producers in Silicon Valley and elsewhere. The varied social structures, culture, and institutions of these economies shape the capabilities of local producers and regional trajectories in unanticipated ways. Cultural differences can, for example, become a source of competitive advantage. Taiwanese companies have significantly shortened product introduction lead-times in part because the strong social networks and personal relationships within a small and closely knit technical community facilitate open information sharing and collaboration across firm boundaries.

Taiwanese firms today are also taking advantage of their understanding of how to do business in China. This is not simply about language and culture, but also about understanding how to get things done. In the words of the CEO of a leading network equipment manufacturer, "Westerners tend to think linearly. Do X, Y, and Z, and a given result must happen. We understand that Chinese societies still very much operate under the principle of 'Nothing is easy, but anything is possible!' We know how to walk a document past dozens of bureaucrats and how to make connections with government officials."[25]

Development on the Ground

A cross-regional community can materialize with great speed, but it typically follows a longer period of relationship-building that is invisible to those outside the community. In the early stages, only a few first-generation immigrants return home to start new businesses, open a branch of a U.S.-based company, or provide venture capital and other services with the hope of exploiting their connections to their home country. Over time, their experiences provide a powerful mechanism for identifying and constructing relationships with home-country counterparts. This is undoubtedly easier in small countries like Taiwan and Israel, but it clearly occurs in India and China as well.

The relationships within a technical community accelerate the identification of promising investments or collaborators. While cross-regional entrepreneurs may initially seek the advantage provided by relatively low-cost technical skill or access to local markets, they quickly recognize the need for differentiation and are forced to deepen local partnerships and participate more actively in local governance. Returning entrepreneurs and managers often work with policymakers to eliminate bureaucratic obstacles, expand higher education, or redefine the curricula of local training programs. Over time this dynamic contributes to the deepening of local capabilities, making local specialists more attractive as partners and forcing competitors to specialize further and upgrade their own capabilities.

The rising levels of workforce skill and capability in the new economies are reflected in the performance by businesses in these countries on global quality benchmarks. The leading Indian software services companies consistently outperform their U.S. counterparts in the Capability Maturity Model (CMM) ranking, developed at Carnegie Mellon's Software Engineering Institute (SEI): virtually all of the Indian companies operate at Level 5, the highest level of expertise, compared to U.S. operations that are more likely to operate at Level 2 or 3. Taiwanese and Chinese manufacturers have similarly achieved top rankings in the ISO9000 standards. While these economies may not be poised to develop leading-edge technologies or applications, they have excelled through "learning by doing" in production.

Firms in Taiwan and India have pursued the incremental but cumulatively significant improvements in production processes and problem solving that now provide a distinctive competitive advantage. These benchmarking standards not only measure and recognize the achievement of world-class capabilities; they also signal a company's excellence to potential partners overseas. The new cross-border start-ups are increasingly able to leverage technical and process capabilities in distant regions alongside the technological and market advantages of Silicon Valley and the U.S. legal system and capital markets.

THE SILICON VALLEY–ISRAEL CONNECTION: BUILDING ON THE MILITARY

Israel's civilian technology sector emerged from a technological environment characterized by decades of investments in higher education and research, a long tradition of applied research and development, and extensive support from a large military-industrial complex. By 1980 the Israeli economy was distinguished by a significant base of knowledge and skill in electronics, optics and communications, and related fields.[26] However, the country generally lacked the managerial and market knowledge needed to commercialize these technologies. In the 1990s the Israeli economy experienced a dramatic transformation and emerged as one of the world's leading centers of high-technology entrepreneurship and innovation.

Israel's Argonauts Set Sail

Most explanations of Israel's technology growth follow the classic model, focusing on military-industrial investments in research and development and a variety of government policy measures designed to promote technology industries: tax incentives, government grants and funding of R&D, training grants, incubators for start-ups, and support for venture capital.[27] But these policies were not that different from those found elsewhere. None of the traditional accounts can explain the timing of the take-off in the 1990s, nor why other, more advanced industrialized nations that boasted well-developed technical education and research capabilities, such as Germany and

France, failed to develop the entrepreneurial and technological dynamism that characterizes Israel today. Long-term investments in education and research were an important and necessary precondition, but explanations that focus exclusively on the domestic political and economic context are not sufficient to explain Israel's superior performance.

The missing variable is the connection to technology centers in the United States. In the 1970s and 1980s, talented Israelis, like their Asian counterparts, came to the United States by the thousands for graduate engineering education. A poor economy and the absence of professional opportunities at home meant that a large proportion of these highly skilled workers stayed and worked in the United States after graduation. Between 1978 and 2000, more than 14,000 professional and technical workers emigrated from Israel to the United States—a sizable proportion of the skill base of a country with a population of 6.4 million.[28]

This brain drain created a network of Israeli engineers and scientists in the United States with unusually close ties to Israel. Israeli engineers in the 1970s and 1980s began traveling between technology regions in the United States and Tel Aviv or Haifa. The strong shared identities and commitment to the survival of Israel motivated many U.S.-educated Israeli engineers to return home and led Jewish Americans to seek ways to contribute to the economy. Today some 500,000 Israelis are living in the United States, and a large proportion have settled in California.[29] Fred Adler, a successful American venture capitalist, reports that his commitment to helping Israel grew out of his first trip there in the early 1970s, when he was dismayed by the poverty of the country. Although he had no prior involvement in Israel or in Jewish activities in the United States, Adler was motivated to devote time and resources to building the country's high-technology industries.

The first foreign technology investments in Israel also occurred in response to Israelis' desire to return home. Intel and National Semiconductor set up IC design centers in Israel in the 1970s in order to retain highly valued engineers. IBM and the semiconductor equipment firm Kulicke & Soffa also established R&D operations in the country. In the 1980s and early 1990s, U.S. multinationals in both the software and semiconductor industries, including Motorola, DEC, KLA-Tencor, and Microsoft, established R&D centers in Israel. These companies were attracted primarily by the high quality of the technical workforce; the Israeli market was small and labor costs were not a concern for these R&D jobs, although the availability of relatively low-cost skill may well have influenced the decision by many of these firms to move from R&D into manufacturing during the 1980s. Survey findings confirm that a majority of the U.S. firms that established R&D centers in Israel in the 1970s and 1980s did so primarily because of pressure from Israeli champions within their company.[30] In most cases, these units have flourished: both IBM and Intel now have their largest non-U.S. research labs in Haifa. Intel has 5,000 employees in Israel, and Motorola employs 4,000. These units have been responsible for key technologies and products within their firms.

In Israel these companies faced the challenges associated with coordination work in teams with very differing communication styles—a challenge that would recur in its other locations, from Taiwan to India. According to the manager of Intel's Israel Microprocessor Division, "There is a huge difference in culture. In Hebrew you can say 'You are wrong.' You can't say that in American . . . We don't have a concept that is part of American culture: Following the rules. If you say to an Israeli, 'Do this,' he will say, 'Why?'"[31] These differences in culture and expectations remain a key challenge in cross-regional collaborations, in spite of sophisticated software tools and processes.

The venture capital and high-technology industries in Israel grew out of common American roots.[32] Fred Adler invested in an Israel-based medical imaging venture called Elscint, started by Dan

Tolkowsky and Uzia Galil in the early 1970s. Elscint's subsequent IPO on NASDAQ was the first for an Israel-based company, and its success depended on Adler's reputation in the U.S. financial community. In 1985 Adler, Tolkowsky, and his son Gideon (who had earned an MBA at Wharton) started Israel's first U.S.-style venture capital fund, Athena, in an attempt to jump-start Israel's high-technology sector.

In the 1960s and 1970s the Israeli ties were predominantly with the east coast, largely because Route 128 was the dominant U.S. technology center at the time. By the 1980s, however, this situation had changed, with Silicon Valley playing a greater and greater role.[33] Athena's charter called for half of its capital to be invested in Israel and half in the United States; this included Israeli entrepreneurs in the United States as well as returnees from the United States. Fred Adler notes: "The turning point was in the 1970s and 1980s when loyal Israeli engineers started going back and forth between the U.S. and Israel, first from the East and then later from Silicon Valley. . . . These interactions were invaluable because they brought an entrepreneurial spirit as well as management talent. The example of companies like Scitex created role models and a demonstration effect for other Israelis."[34]

One of Israel's first return entrepreneurs, Aryeh Finegold, had worked for Intel in Silicon Valley in the 1970s before starting Daisy Systems, a computer-aided engineering company, with Fred Adler as the lead investor and chairman of the board. When Daisy Systems failed during the 1980s, Finegold returned to Israel. With funding from Athena he started Mercury Interactive, an Internet software testing and monitoring company. Finegold was not alone: he recruited Amnon Landan, a co-worker from Daisy Systems, to return to Israel and join the founding team. Landan is now the company's CEO and was named "Entrepreneur of the Year" in 2003 by *Forbes* magazine.

Mercury Interactive is a classic cross-regional business: its origins, leadership, and virtually all of its employees are Israeli, yet the company is headquartered in Silicon Valley, incorporated in Delaware, and listed on NASDAQ, having completed a public offering in 1993. As of 2002, it had a market capitalization of over $4 billion. Edward B. Rock calls it a "Yankee" firm that has been made to "look, sound, and feel like a U.S. firm" even though its origins and top leadership are Israeli.[35]

Supporting Success

The practice of bypassing the Tel Aviv Stock Exchange (TASE) and relying on U.S. capital markets for liquidity has become standard for Israeli start-ups.[36] It originated with Fred Adler and Athena and continues partly as a result of the extensive participation of U.S. venture capitalists and investment bankers, and in part because of the structure of the Israeli economy, where a small number of major families control over half of the capital on the TASE. This contrasts with the situation in Taiwan, where the venture capital model was also imported from Silicon Valley by overseas Chinese, but the vast majority of domestic companies have gone public on the Taiwan Stock Exchange.

These ties to the United States and the creation of an indigenous venture capital industry created an infrastructure in Israel that supported technology entrepreneurship in the 1990s. In 1991, only one venture capital firm was operating in Israel; by 2000 more than a hundred such firms were active, and they invested over $1 billion per year in start-ups. This rate of investment, which returned to pre-recession levels in 2003, made Israel the world's third largest recipient of venture capital, following only Silicon Valley and Boston's Route 128 region.

Successful firms like Mercury Interactive proliferated in Israel in the 1990s, and the infrastructure of professional service providers expanded to include not only more venture capitalists but also lawyers, accountants, investment bankers, and consultants. Government agencies also sought to support entrepreneurs through programs like Yozma, a public venture capital fund. And the immigration of an estimated 100,000 scientists and engineers, mainly from the former Soviet Union following the fall of the Berlin Wall, deepened Israel's supply of technical skill. This mass migration ensured a short-lived wage differential for engineers of about 30 percent vis-à-vis the United States, but by the late 1990s Israel, like Silicon Valley, was facing labor shortages.[37] These newer immigrants typically lacked a market orientation or even experience with modern technology, so very few became entrepreneurs.

Connections to the United States expanded as well. Silicon Valley–based firms such as Lightspeed Venture Partners, Sequoia Capital, and Benchmark Capital began investing in Israeli start-ups in the 1990s. Between 2000 and 2003, Intel Capital alone made thirty-five investments in Israeli ventures, and more than half of the venture capital raised by Israeli companies since 2000 came from foreign investors.[38] The San Francisco–based venture capital firm Blumberg Capital provides assistance to Israeli companies seeking to open offices in Silicon Valley, in addition to investing in early-stage IT companies based in either region. And the Israel Economic Consulate in the San Francisco Bay Area has promoted cross-pollination between the United States and Israel by setting up introductions for entrepreneurs and companies.

Itzhak Levit, CEO of CDI Systems and a retired lieutenant colonel of the Israeli Defense Force, describes the results of this cross-fertilization: "Why were there so many entrepreneurs in Israel in the 1990s? One-quarter of my university graduating class went to the United States and then stayed on to work in high-tech in Silicon Valley. They all started coming back to become entrepreneurs. . . . They knew how to hire U.S. marketers and business developers."[39]

Going Critical

A critical turning point in the development of an entrepreneurial system occurs when capital generated by local start-ups is reinvested in a new generation of start-ups. At this point, the local system of entrepreneurship gains the potential to become self-sustaining: local knowledge begins to accumulate, lessons and experience are shared across firms and generations, and repeat entrepreneurship becomes both common and acceptable. In 2004 local observers estimated that as many as 30 percent of Israeli technology start-ups were headed by experienced entrepreneurs.[40]

One could argue that the turning point was reached in Israel when Mirabilis, which created ICQ chat software, was acquired in 1998 by America Online for $407 million. Co-founder Yair Goldfinger, who owned 21 percent of the firm, went on to invest in a range of local technology start-ups. Although the team lacked extensive American experience, they moved to San Jose to develop their product; this may have influenced their decision to give the ICQ software away for free, creating the model of viral marketing (marketing by word of mouth that exploits network externalities). Goldfinger is active in several start-ups: he talks with founders or visits the companies weekly, and offers both business and technical assistance. He is also serving as the chief technology officer of a start-up that he helped found. Mirabilis served as an important role model that inspired a younger generation of Israeli software engineers and programmers to become entrepreneurs. Goldfinger reports: "Mirabilis helped change the high-tech industry and inject new blood in the system. It was very inspiring. Lots of kids saw what we had done and decided they wanted to be the next Mirabilis."[41] A study of Israel's high-tech sector confirms this emergence of indigenous entrepreneurship in the 1990s:

> Virtually all of the earlier high-tech entrepreneurs in Israel had spent time in the United States. . . . First-hand exposure to the high-tech environments in the U.S. was extremely important for showing what was possible, as well as for providing would-be entrepreneurs with contacts and market knowledge . . . [it was] not until the 1990s that people with no experience abroad or in a multinational firm began to found companies in significant numbers.[42]

The other indication of the growing internal dynamism of the Israeli technology system is the dramatic decline in the emigration of native Israelis to the United States during the 1990s.[43] (A comparable and equally abrupt drop-off in emigration among skilled Taiwanese graduates during the same period suggests that the professional opportunities in home economies had become so attractive that young people who a decade earlier would have left to study in the United States increasingly chose to remain at home.) This does not mean, however, that Israel's ties to the United States have disappeared. Israeli companies still depend heavily on a presence in Silicon Valley. The early and strong connections to the United States oriented Israeli entrepreneurs toward this country and the most sophisticated technology market in the world rather than to closer European markets. Eli Barkat, CEO of the Internet software company BackWeb Technologies, explains that presence in the United States is essential to getting to the market fast. Barkat started the firm in Israel but subsequently, realizing the need to be close to customers, he opened an office in New York. However, even New York was not close enough: "I found myself flying once a week to the Valley. . . . Then I started staying two days and sleeping a night. And then I found myself spending most of the week here and flying back to New York to sleep."[44] Finally, he moved his U.S. office to San Jose. The firm is now split between the Valley, where the sales and marketing employees are based, and Israel, where engineers work on product development.

Evidence of this phenomenon is provided by the data on U.S. visas, which suggests that while permanent emigration of Israelis has declined, the short-term movement of highly skilled individuals continued or even increased during the 1990s. The chairman of VDONet, another Israeli company, explains: "Theoretically the Internet is global and you can sell anything from anywhere. But to have [a] serious presence with major customers, it's still done the old way of having a chat over breakfast, lunch, or dinner. The Internet hasn't changed that." And David Lavenda, vice president of marketing for Business Layers, says simply: "I spend half my time in the U.S., half my time in Israel, and the other half on the airplane in between."[45]

Tremendous amounts of information and know-how are transferred as engineers and entrepreneurs travel between Silicon Valley and Israel, yet there is a minimum of institutionalization. In contrast with the mechanisms for information exchange within a large hierarchical corporation, information travels through the decentralized and informal networks created by venture capitalists, service providers, consultants, and suppliers.

Self-Organization with Military Roots

In contrast to the Chinese and Indian communities, the Israeli community in Silicon Valley has flourished with very little organization. A network of Israeli investors and executives in the region met informally during the 1990s, and the Israeli consulate in San Francisco continued to build connections. In 1998 the Carmel Dalia Organization was created as a local alumni group of the Technion-Israel Instituite of Technology; and in 2002 the SIVAN Group was organized to provide a forum for Israeli executives to "network with their peers, learn from each other, and assist the Israeli high-tech community in Israel and the Valley." SIVAN is a Hebrew name and also an acronym for Silicon Israeli Valley American Networking.

The community has been drawn together for more than a decade primarily by the shared identities and strong social bonds created from service in the Israeli Defense Force, particularly the elite technical units that provide training in advanced technologies like computing and communications—including the Air Force, the Computing Center, and the Intelligence Corps. Four to five years of service in these elite military units, according to Israeli technologists, leads to a dense, close-knit community that does not have a need for more formal organization. In addition to socializing with each other, Israeli families share information, help one another with recruitment and financing, and start companies together. One observer compares it to a fraternity.

A disproportionate share of Israel's entrepreneurs come from these elite military units, and engineers who served together in the same unit frequently start companies together. Nir Barkat, a former major in the Israeli Paratroops Corps and co-founder of BRM Technologies, notes: "Once our role models were soldiers. . . . Now they're entrepreneurs."[46] In the words of retired Lieutenant Colonel Itzhak Levit, "The 10 percent of the army's elite are the ones that become IT managers and entrepreneurs; two-thirds of Israeli entrepreneurs have military backgrounds because the education provides the tools to think strategically, to work under uncertain conditions, and to operate with limited resources. This trains them to take risk and solve problems fast."[47]

The strong social bonds that carry over from the military also create dense networks for problem solving and information exchange. One alumnus of the elite military units asserts: "I am not afraid to take any consulting job in the field because I know that in the worst case scenario I am at most four phone calls away from a world expert for any software-related problem."[48]

The military has contributed to the development of Israel's technology growth in many other ways, from providing advanced technical training and developing teamwork and leadership skills to serving as a sophisticated market and diffusing technology and know-how to the civilian sector. According to Yaacov Zerem, co-founder of Ophir Optronics, a world leader in laser measurement devices and optics:

> It was the military's high level of requirements that enabled us later to develop the complete line of products that we now sell to the civilian sector. The military's driving us to develop state-of-the-art equipment forced us to improve our R&D and production standards. We were able to cooperate with the IDV—they would keep us informed about upcoming projects so we could develop optics for the future. . . . I believe this is why the defense establishment has been such an important driver for high-tech here.[49]

Finding a Niche

Israel is now home to some 4,500 technology companies as well as more than 300 venture capital funds and investment companies. There is a clear pattern of niche specialization in the country's technology sector. Many companies have specialized in Internet software, particularly security and communications-related software—which is not surprising, given the focus of military research. This includes well-known companies like Check Point Software (Internet and network security software) and Mirabilis (ICQ instant messaging software) as well as less well-known companies like Business Layers (enterprise software) and Virtual Domination (virtual studios).

The pattern of niche specializations in Israel's technology enterprises suggests that Israel and Silicon Valley cannot be viewed as competitors (although, of course, individual firms may compete fiercely); rather, the two are largely complementary. Together they are more innovative than either one could be alone. Nor is it possible to explain the interregional linkages solely on the basis of the internal structure of multinational firms, since so many other business and social links bind Israel and Silicon Valley together and shape the possibilities for development.[50]

NATIONAL POLICY AND CROSS-REGIONAL ENTREPRENEURSHIP

As we have seen, immigrant technical communities are accelerating the cross-regional circulation of information, business models, and technical know-how. While return entrepreneurs and investors can bring home the technical and organizational know-how associated with technology entrepreneurship as well as their networks of contacts, their impact on their home-country economy depends as much on domestic institutions and resources as it does on their own expertise and relationships. Despite their shared personal connections to Silicon Valley and elements of its decentralized industrial system, the new technology regions are developing distinctive specializations and capabilities. Differences in size, demography, history, and geopolitical context as well as political and economic institutions help account for the divergent regional trajectories.

The developmental trajectories of Israel and Taiwan are those of relatively small and open economies with close political and economic ties to the United States. Both have benefited from postwar military alliances with the United States and the resulting technology transfers, and the United States has been a preferred destination for the education of senior bureaucrats as well as engineers. Policymakers in both nations invested heavily in higher education and in research and development, which were essential preconditions for their successes. However, the ability to transform research findings into commercially viable innovations requires market knowledge and connections as well as technology and skill. In both places this was achieved through a bottom-up process of entrepreneurial experimentation—and the long-distance connections to Silicon Valley through decentralized networks of immigrant engineers, entrepreneurs, investors, and managers.

In both Taiwan and Israel, the small size of the economy has allowed the overseas technical community to work closely with local policymakers as well as with educators, researchers, entrepreneurs, managers, and investors to reshape local institutions. As in Silicon Valley, open communication among these groups has facilitated the definition of policies that support technology experimentation and entrepreneurship-led growth. As a result, the technology industries in both countries have emerged through a complex interplay of public policy and private initiative. Neither can be described as state-led, planned, or completely organic.[51] The mobilization of the resources of an overseas technical community in the service of economic development required policymakers and returnees to develop a shared language and vision that allowed them to jointly identify and overcome obstacles to entrepreneurship-led growth. Although both Israel and Taiwan began with a technically skilled workforce and a well-functioning physical and communications infrastructure, they had to develop institutions to support venture capital funding and provide viable liquidity options, research and training capabilities, collaboration between university researchers and private entrepreneurs, and openness to foreign investment and trade.

These economies both rely heavily on export markets, and the transnational community has played a central role in connecting start-ups with fast-changing technology and market trends as well as with leading customers. During the 1990s technology entrepreneurs in Taiwan and Israel defined and grew new market segments, rather than competing head-on in markets that U.S. producers already dominated. Their experience working in Silicon Valley, in particular, was essential to their ability to identify and gain access to underserved markets, recognize and follow emerging market standards, and define niches that are appropriate for new start-ups.

Taiwan and Israel have also developed sizable domestic markets for their products, enhancing their ability to innovate through interaction with local customers. For example, Taiwan's PC and systems firms are important customers for the local semiconductor industry as well as for its many components manufacturers. The Israeli military served as a technically demanding customer and driver of high-tech research and development. Both cases demonstrate the capacity for localized upgrading in a decentralized and entrepreneurial industrial system as well as the advantages of connections to global markets and technology trends.

The development of innovative capabilities in both Israel and Taiwan is well documented. Trajtenberg shows that between 1970 and 1998 both nations accumulated U.S. patents at a substantially higher rate per capita than their counterparts at comparable levels of development, including Ireland, Spain, and New Zealand as well as Singapore, Hong Kong, and South Korea.[52] By 1998 they had surpassed even wealthier economies such as France, Italy, and the United Kingdom in per-capita patenting. These patents were concentrated in leading-edge technologies including electronics, computers, and communications, and the proportion granted to domestic firms and institutions grew rapidly in the 1980s and 1990s. Only 5 percent of Taiwan's and 17 percent of Israel's patents during this period were assigned to foreign-owned firms and institutions, in contrast to economies dominated by multinational investments such as Singapore (56 percent to foreign-owned companies) and Ireland (44 percent).

Taiwan adopted a very different development strategy in the technology industries during the 1970s and 1980s than its East Asian counterparts. Government policy has supported entrepreneurial ventures, and investment in the technology sector is overwhelmingly from domestic private-sector sources (83 percent), with only 5 percent from the government and 12 percent from foreign investors (many of whom are Taiwanese). In Korea and Japan, by contrast, the government leads the economy in alliance with a handful of very large, diversified corporate groups. Capital markets are structured to privilege the large state-supported corporations, the *chaebol* in Korea and the *keiretsu* in Japan, and to squeeze out entrepreneurial or small competitors. The most talented youth in these countries aspire to high-status jobs in the government bureaucracy or leading firms rather than entrepreneurship. Singapore is similar in that the government controls the economy tightly, with state and linked businesses creating 60 percent of GDP; however, it has aggressively opened up to foreign investment, with over 80 percent of the manufacturing investment in Singapore coming from multinational firms.[53]

These cases suggest that the nation-state and the large corporation are adapting to—rather than defining or controlling—an environment in which decentralized networks of entrepreneurs, engineers, financiers, and other service providers mobilize resources rapidly to respond to unpredictable markets and technologies. The pronounced localization of technology entrepreneurship in the Tel Aviv–Jerusalem region of Israel and the Hsinchu–Taipei corridor in Taiwan also underscores the importance of the collocation of highly specialized skill and capabilities as an enduring source of innovative capability and competitive advantage.

The importance of the regional scale for information exchange is illustrated by the rankings of Ireland, Taiwan, and Israel as number 1, 2, and 3 respectively in the 2002 Wealth of Nations Triangle Index. The Index, formulated by a nonprofit development organization, aggregates sixty-three indictors of a nation's social environment, information exchange, and economic environment. These three small nations are distinguished in the pool of seventy countries by their top rankings in information exchange—which includes measures of the nation's information aptitude (literacy rate, tertiary education, and so on)—as well as the development of information infrastructure and information distribution. They also rank number 1 (Ireland), number 2 (Taiwan), and number 8 (Israel) for their economic environment, which includes standard measures of national economic performance as well as internationalization of the economy and the business environment.

India and China, both very large developing economies that opened their markets to the global economy in the last two decades, have less clearly defined trajectories. The political and economic environment in these countries is complex, multilayered, and regionally differentiated. Both countries have attracted foreign investment targeted at their low-cost, technically skilled workforces. For both, brain circulation with an overseas community in Silicon Valley has been an important factor in fueling the rise of technology entrepreneurship. National policymakers from both countries have established ties to their large technical communities in Silicon Valley, with local and regional governments in each country competing intensely for technology activity. And in both China and India, technology investment and entrepreneurship remain concentrated in a handful of urban centers, in spite of rapidly rising costs.

The leading Indian technology firms today specialize in exporting software services to American and European corporations and are concentrated in a few large urban centers in southern India, including Bangalore and Hyderabad. The leading Chinese IT businesses focus on manufacturing for the domestic market, and are concentrated in the leading metropolitan areas along the east coast of

China, particularly Beijing, Shanghai, and Guangzhou. China has benefited as well from growing ties to Taiwanese businesses, despite the political tensions across the Straits. These regions are now home to the kinds of cross-regional investments that promise to increase competitiveness in established markets or to grow sizable new markets. However, it is still early in the process: neither China nor India has seen a full cycle of entrepreneurial success and reinvestment.

Some differences can be traced to institutional factors. In China, governments at all levels have channeled substantial investment into developing an urban infrastructure and growing the domestic market. The selection of IT industries as a national priority has also triggered increased investments for research and development, technical education, the adoption of information technology in both public and private sectors, and the construction of a seemingly unlimited number of science parks. However, political control by the Communist Party, ongoing corruption, and chronic weaknesses in financial markets continue to create uncertainties about China's future.

India, by contrast, made expansion of export earnings a top priority in order to improve its balance of payments at a time when distrust of bureaucratic intervention (based primarily on the failures of the postwar "license Raj") led to the gradual embrace of liberalization and a largely hands-off approach to industrial development. Nonetheless, regulatory obstacles remain significant, and progress toward improving the domestic physical and telecommunications infrastructure has been slow.

Both India and China are increasingly important actors in the global economy, particularly in technology industries. There are, however, many unanswered questions about their developmental trajectories. Technology is a small proportion of overall economic activity in both countries, and it remains highly geographically concentrated. Still, it could have a disproportionately large impact over the long term. For both China and India, as the experience of Israel and

Tawain suggests, much will depend upon the willingness of domestic political and business leaders to collaborate with returning engineers and entrepreneurs in developing a shared vision of the economic future—including strategies for removing institutional and political obstacles to entrepreneurship, and for ensuring investment to upgrade domestic technological infrastructure and capabilities.

4 ❖ Taiwan as Silicon Sibling

Ronald Chwang was one of Taiwan's first cross-regional entrepreneurs. Born in China and raised in Taiwan, Chwang received a Ph.D. in electrical engineering from the University of Southern California in 1975. He worked for technology companies in North America until 1983, when he was recruited away from Intel to become chief engineer for one of Taiwan's first semiconductor enterprises, Quasel. At Quasel he established a Silicon Valley R&D center to complement the firm's manufacturing in Taiwan. In the early 1980s few overseas Chinese engineers had returned to Taiwan because the economy was still very poor. In Chwang's words, this meant that "when I worked at Quasel USA I lived between Silicon Valley and Taiwan. I was always on airplanes."

Quasel failed in 1986, but Acer CEO Stan Shih—who knew Chwang because Acer was an investor in Quasel—recruited Chwang to run Acer R&D Labs in Taipei, where he spent five years before returning to Silicon Valley. In 1991 Chwang took over a recently acquired business in San Jose that became Acer America. As president and CEO, he set up and ran a sophisticated PC manufacturing center and returned to Taiwan regularly for board meetings. At the same time, Chwang was active in the local branch of the Chinese Institute of Engineers and in the formation of the Monte Jade Science and Technology Association.

In 1997 Chwang became chairman and president of the newly formed Acer Technology Ventures, America (ATV). ATV invests in technology start-ups on both sides of the Pacific and "seeks to build partnerships with entrepreneurs, combining extensive Acer Group contacts and capital resources with managerial, marketing, and technical experience to create value for young start-ups." ATV has offices in San Jose, Taipei, Shanghai, and Beijing, and the majority of its investments are in either Silicon Valley or Taiwan (although a growing number are located in China). Most of its portfolio firms have Chinese founders or senior managers.

Ronald Chwang's career is not unusual in Silicon Valley. He has experience in both start-ups and more established technology companies, he has worked as an engineer, manager, executive, and venture investor, and he has been involved with a range of industries from semiconductors and personal computers to peripherals and servers. As he changes jobs, he expands his professional networks and reputation; he also transfers skill and know-how across both firms and business units.

Chwang's career is distinctive, however, because it bridges two distant regional economies. He has the language skills and the institutional and cultural know-how to operate effectively in both Taiwanese and U.S. business environments, and he maintains relationships with engineers and entrepreneurs in both Silicon Valley and Taiwan (and increasingly in mainland China as well). Yet his career also differs dramatically from that of the traditional multinational manager. Chwang has worked for Taiwan's largest IT firm for more than a decade, yet his activities extend well beyond the corporate boundaries of the Acer Group. Acer's open "cellular" structure encourages him to integrate into the technical and professional networks in Taiwan and the United States; his ability to do so enhances his contributions to Acer, whether they involve building partnerships with suppliers and customers, collaborating with other investors, serving on the board of start-ups or public companies, advising policymakers in Taiwan, or mentoring entrepreneurs in Silicon Valley's Chinese community. In short, Chwang's career illustrates how even managers of large corporations can benefit from the local industrial infrastructures of the Silicon Valley and Hsinchu economies.

THE TAIWAN NETWORK

Ronald Chwang is part of the technical community that transformed Taiwan from a poor island specializing in producing toys, umbrellas, and footwear into a global center of electronic systems design, manufacturing, and logistics. In the late 1970s, U.S. and Japanese consumer electronics and semiconductor corporations looked to Taiwan for low-cost labor. The resulting foreign investment led to a large increase in the number of indigenous Taiwanese producers in the 1970s and 1980s, many of which began to export components for personal computers. The growth, diversification, and improvement of this infrastructure, along with the development of integrated circuit (IC) manufacturing capabilities, allowed Taiwanese producers to gain a growing share of global IT manufacturing in the 1990s. This growth came at the expense of other comparably positioned Asian competitors such as Singapore, Malaysia, and Korea. By the end of the decade Taiwan ranked as the world's third largest producer of IT hardware with over $20 billion in output, following only the United States and Japan. Thus in just over a decade, IT exports grew to account for close to half of the island's total exports and led to a doubling of Taiwanese per capita income—from $6,333 in 1988 to $13,235 in 1999.[1]

Taiwan's growing technological capabilities were reflected in its rapid rise in the international rankings of U.S. patent recipients: in 1980 Taiwan ranked twenty-first; by 1990 it had reached number 11; and in 1995 it was seventh.[2] As Figure 4.1 demonstrates, since

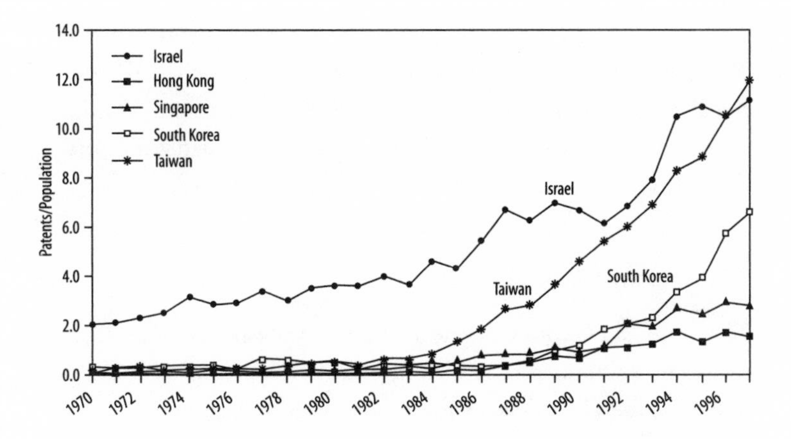

Figure 4.1. Patents per capita for Taiwan, Israel, and the newly industrialized countries (patents per 100,000 population). *Source:* Trajtenberg 2000.

the mid-1980s Taiwan has received substantially more patents per capita than the other newly industrialized countries in Asia, including South Korea, Singapore, Malaysia, and Hong Kong. Along with Israel, Taiwan now ranks ahead of all the G7 nations except the United States and Japan.[3] Other indicators of Taiwan's technological capability, such as papers in leading engineering journals, tell a similar story. According to the NCTU Web site, National Chiao Tung University in the late 1990s contributed more papers to the top IEEE journals than leading U.S. universities and corporations such as Stanford, MIT, and IBM, and National Taiwan University ranked very high as well.

How did a small and (until recently) poor island of 24 million people surpass other Asian economies as well as more advanced economies in global technology competition? The central but largely unrecognized actors in Taiwan's accelerated industrial upgrading are U.S.-educated engineers and entrepreneurs like Ronald Chwang, who built a social and economic bridge linking Silicon Valley and Taiwan. These former classmates and colleagues, who shared educational backgrounds and professional as well as cultural identities, were crucial to the successes of more commonly recognized actors—the state and large corporations.

Assessing the Classical Model

The dominant accounts of the Taiwanese postwar economic miracle focus on the state's role in directing the development of traditional industries like chemicals, textiles, and electrical machinery.[4] Students of Taiwan's PC and IC industries attribute its successes in technology to the institutions of a developmental state such as the publicly funded research and development agency, Industrial Technology and Research Institute (ITRI), and its subsidiary, the Electronics Research and Services Organization (ERSO).[5] Amsden and Chu credit these institutions with ensuring the timely transfer of technology to domestic firms that mass-manufacture products based on technologies developed elsewhere, while also financing industrial diversification into new areas of production.[6] In this view, the state in Taiwan supports the upgrading of large, established corporations as they relentlessly lower costs through the rapid ramp-up and achievement of scale economies in mature product lines.

This approach, which interprets the Taiwanese experience through the conceptual lenses developed to explain late industrialization elsewhere, overlooks the distinctive organization of production in Taiwan, which provides a competitive advantage in an environment of market volatility and technological uncertainty. In contrast to the giant diversified corporations that dominate the IT industries in Japan and Korea in close alliance with the state, Taiwan's IT industry consists of some 10,000 highly specialized producers, mainly small- and medium-sized companies, within a fifty-mile region known as

Hsinchu, located between Taipei and the Hsinchu Science Park in northwest Taiwan. The state has actively supported this technology industry, generally by encouraging the entry of new competitors rather than by supporting leading firms.

Scholars have recently recognized the importance of Taiwan's integration into global supply chains, which consist of networks of independent producers in developing economies that manufacture components for branded retailers in the advanced economies.[7] The fragmentation of IT production facilitates the geographic relocation of manufacturing to low-cost locations. Taiwan, for example, entered global technology markets through original equipment manufacturing (OEM) relationships with PC retailers. An OEM arrangement is one in which the customer provides detailed technical blueprints and components that allow the contractor to produce a product that will be marketed under the customer's brand name.

However, by characterizing these OEM relationships as hierarchical and dominated by "global flagship producers," this view overlooks the continuing engineering improvements and process innovations at all levels of the supply chain in Taiwan. While relationships with OEM customers were certainly a source of technology and know-how for Taiwanese producers in the 1980s, they contributed in subsequent decades to the building of local capability independent of these relationships.

The Hsinchu economy today bears a striking resemblance to that of Silicon Valley: its producers are clustered in a small geographic area, and the economy is characterized by high rates of new firm formation and failure. Hsinchu's thousands of specialist component producers jointly constitute a well-developed and highly responsive design and manufacturing supply chain for PCs, cell phones, digital cameras, and related electronic appliances and systems.

The competitiveness of this industrial system lies in the ability of local producers to continually and quickly improve product quality and performance while also reducing costs. This means that the economy's dynamism cannot be well understood at the level of the individual firm. Even as firms grow, they still depend on other specialist producers—both local and global—and they benefit from widespread incremental improvements at all levels of the supply chain. Local industrial leaders have long recognized this well-developed component supply infrastructure as their key advantage in a fast-paced and uncertain industry. As C. S. Ho, president of the Mitac Group, put it in 1993:

> Time to mass production is critical. Taiwan has some cost advantages over U.S. firms, but the real key is that we can act really fast. . . . Parts can be in the critical path to fast ramp-up. It doesn't matter how small the part is, if it's not there, you can't build the machine. . . . [In Taiwan there are] lots of companies doing different things, lots of companies doing the same thing. The result is a virtuous and intensively competitive cycle that drives speed and innovation.[8]

Historically, Taiwan's producers have depended on outside providers of high-end microprocessors, hard disk drives, and some memory chips, and most have lacked the distribution and marketing capabilities that would allow them to successfully develop their own branded products. However, this has not impeded their ability to continue introducing new and differentiated components and end-products. Taiwan-based firms have continued to gain both technological and operational capabilities; today they design and produce leading-edge electronic systems, from Dell laptops and Sony Playstations to Apple iPods, Motorola handsets, and Palm PDAs.

Network-Based Agility

Taiwan's industrial system has enabled local producers to integrate new technologies and shift to new products through low-cost engineering and manufacturing. By the late 1990s, Taiwanese companies served a diverse range of markets including automotive equipment, telecommunications, industrial computers, and consumer electronics.

Taiwan lost its original labor cost advantage over the rest of Asia in the late 1980s, yet the growth of its technology industry accelerated. The widespread shift in the 1990s from original equipment manufacturing (OEM) to original design manufacturing (ODM) relationships—in which the manufacturing partner takes responsibility for design as well as engineering, production (including component procurement), inventory management, and logistics—underscores the increasing sophistication of local capabilities and the growing autonomy of local producers. The Hsinchu region is now home not only to world-class IC foundries and super-efficient PC manufacturers, but also to some of the world's most advanced suppliers of apparently mundane components like computer monitors, motherboards, connectors, keyboards, scanners, modems, power supplies, and displays.

There are now more than twenty motherboard companies in Taiwan, ranging from small firms making niche-market motherboards to first-tier manufacturers like Foxconn Electronics (the registered trade name of Hon Hai Precision Industry). In 2004, these firms manufactured 79 percent of the world's motherboards.[9] In the same year Taiwan's Quanta Computer and Compal Electronics, along with about a dozen other firms, produced 70 percent of the world's notebook computers (35–40 million units), primarily for the leading U.S. and Japanese vendors. Many of these same companies, along with others, also produced PDAs (55 percent of worldwide shipments), pocket computers, handsets, and smart phones. Taiwanese firms like AU Optronics, Chi Mei Optoelectronics, and TPV Technology produced over 65 percent of all liquid crystal display (LCD) monitors; others, including AOpen, Behavior Tech Computer, BenQ, and Lite On IT, controlled 45 percent of the world's optical disk drive business; and Microtek alone, a company that started by making scanners, produced 40 percent of world shipments of digital still cameras.

These and hundreds of other relatively unknown firms are partners in the success of the higher-profile companies like Taiwan Semiconductor Manufacturing Corporation (TSMC), Acer, and Quanta. These companies have developed their own capabilities by collaborating with customers and suppliers to develop new products or improve existing products. Even as established producers grow and diversify, other specialists emerge and innovate in new niches, contributing to a deepening division of labor and a diversification of the industrial system.

Scholars who restrict their analytical focus to Taiwan's cost advantages, the role of the state, large corporations, or global production networks often fail to recognize how this decentralized organization of production allowed Taiwanese producers to surpass their Southeast and East Asian competitors. Local producers have drawn on technological resources and talent from Silicon Valley and elsewhere in the world, as well as from domestic research institutes and universities. This connection to both local and distant expertise has been essential to upgrading the capabilities of the entire system.

National Champions and Industrial Decentralization

In the early 1990s, Levy and Kuo compared the "bootstrap" strategy of Taiwan's small PC-related firms and their propensity for risk-taking and experimentation with the high-volume assembly strategy of Korea's giant *chaebol.*[10] They concluded that Taiwan's decentralized infrastructure produced an ongoing stream of innovation. Korean companies, by contrast, failed to keep up with shorter product life cycles and were increasingly forced to purchase key components like motherboards from Taiwan.[11] In 2004, Taiwan became the world's leading producer of flat-panel displays even though this technologically advanced and capital-intensive manufacturing process would once have been viewed as the sole province of deep-pocketed industrial giants in Korea and Japan.[12]

Jason Hsuan, CEO of Taiwan's TPV Technology, which surpassed Samsung Electronics to become the world's leading supplier of LCD monitors, believes that his firm's competitive advantage lies in its focus (on monitor production alone), its close relationships with its panel suppliers, a diverse range of customers to minimize risk and facilitate learning about markets and product improvements, and a strong R&D workforce to ensure high-quality engineering. Doubtless the firm also benefited from its early entry in China, where it set up a sales office in 1990 and manufacturing in 1992.[13] The firm is now investing heavily in plasma display technology (PDT) for TVs.

Japanese IT producers—in spite of their leading-edge technology and talent—also continue to lose market share to Taiwan and the United States in electronic systems and even consumer electronic products.[14] Although Japan first perfected the efficient mass manufacturing of electronic components to achieve high quality and low cost, captive component suppliers remain largely dependent on traditional *keiretsu* relationships. These relationships not only limit their autonomy but also reduce the incentive to increase efficiency and responsiveness or to lower costs.

Southeast Asian producers were also potential competitors for Taiwan. In the 1960s and 1970s Singapore and Malaysia, like Taiwan, were the preferred destinations for U.S. semiconductor and consumer electronics producers seeking cheap labor. Their trajectories diverged in the 1980s when Singapore aggressively—and successfully—recruited higher value-added foreign investment and became a manufacturing center for U.S. disk drive companies and other high value-added PC component manufacturing. Malaysia followed Singapore's strategy, and by offering still lower costs, it succeeded in attracting investments away from Singapore. However, the IT industries in both countries were controlled by the state and multinational corporations, whose investments in capital-intensive manufacturing branch plants limited the development of local linkages and stifled indigenous entrepreneurship.[15] Penang, Malaysia's oldest and most advanced technology center, still lacks a significant capacity for independent entrepreneurship or innovation.[16]

Taiwan's systematic encouragement of entrepreneurship, the growth of indigenous small- and medium-sized enterprises, and the incremental and cumulative upgrading of domestic capabilities stand in sharp contrast to the experience of Singapore and Malaysia.[17] The manager of a Singapore-based PC firm complained in 1994 about Taiwan's superior expertise in motherboard design:

> If I had my own choice I would do it myself. But we are way behind Taiwan. . . . In fact their design capability exceeds anything in the United States as well. They can fax designs to you that afternoon. For years, they have been building up design cells, so when they need a new design, they pull together the appropriate cells, make some small changes, and they have a new design.[18]

Policymaking Partnership

Taiwan's policymakers do not direct this process of technological and industrial upgrading, and institutions like ITRI and ERSO have limited control over the pace and direction of innovation in domestic industry. However, the public sector has played a crucial role in the governance of the industrial system. Starting in the 1970s local policymakers worked closely with overseas Chinese engineers, developing a policy consensus that diverged both from the existing strategy based on export processing zones and from the then-dominant practice in Asia of publicly financing large-scale producers or "national champions." They sought instead to create an environment to support competition and innovation in technology-related industries by investing in higher education, research and training, benchmarking, and facilitating ongoing exchanges among engineers. An emphasis on entrepreneurship was explicit in Taiwanese policy as early as a 1961–1964 economic plan, which states: "It is true that Taiwan is short of capital, but what are wanted most is entrepreneurs and their entrepreneurship."[19] Tellingly, when establishing financial incentives and subsidies, policymakers designed them to encourage new entrants in a sector rather than assisting incumbent producers. They also introduced laws that enabled venture capital, and reformed their capital markets to support the creation and growth of small, specialized companies.

In the 1970s and 1980s, overseas Chinese engineers in Silicon Valley and in Taiwan simultaneously formed local associations and advised senior Taiwanese ministers and policymakers. These conversations between engineers and other professionals on both sides of the Pacific—a majority of whom had graduated from one of the island's four elite engineering universities (National Taiwan University, National Chiao Tung University, National Tsinghua University, and National ChengKo University)—contributed to the creation of collective identities and commitment to a shared project among a far-flung technical community. U.S.-based engineers provided insights into the changing organization of IT production, and particularly the advantages of specialization in a volatile and uncertain environment. Meanwhile, Taiwan's policymakers devoted public resources to designing policies and institutions to support industrial decentralization.

Taiwan's political leaders in the postwar period were overwhelmingly technocrats, with graduate degrees in engineering-related fields from universities in the United States and Japan. They had been exposed to the world's largest and fastest-growing economies, industries, and companies. These policymakers aggressively promoted programs and institutions to develop Taiwan's technological capabilities. They focused initially on improving the skill base and technical infrastructure in what was still a very poor country: Taiwan's GNP per capita in 1962 was $170 in U.S. dollars, on a par with Zaire and the Congo.[20] The policymakers complained bitterly about the brain drain that occurred from the 1960s onward, when engineers left by the thousands to study in the United States; very few returned home. In fact, Taiwan sent more doctoral candidates in engineering to the United States during the 1980s than any other country, including entire graduating classes from Taiwan's most elite engineering universities. These students were attracted by the fellowship money available for graduate studies at American universities and were pushed to emigrate by the absence of professional opportunities at home. Most stayed in the United States after graduation, recognizing that there would be little demand for their skills in Taiwan.

By implementing policies that supported competition, however, Taiwan's policymakers created an environment that encouraged the eventual return of many of its overseas engineers. Patrick Wang, a Stanford graduate who worked at Hewlett-Packard for almost two decades before returning to Taiwan to start a company, described this role: "We advised government officials on Taiwan's strategy continuously in the 1970s and early 1980s, both formally as consultants and also informally. Over time we developed a shared strategic vision that made it easier to return."[21] Genda Hu—who earned a Ph.D. in electrical engineering from Princeton University and worked in Silicon Valley semiconductor firms for twenty years before returning home to become director general of ERSO in 1996—refers to a process of "cross-breeding" and "mutual learning" between engineers in the two distant industrial systems.[22]

Anchoring the Network

The development of a cross-regional technical community transformed the relationship between Silicon Valley and Hsinchu. In the 1960s and 1970s, capital and technology from the United States and Japan found their way to Taiwan mainly because of multinational corporations seeking cheap labor. This core-periphery relationship had given way by the 1990s to less centralized two-way flows of skill, technology, and capital between the regions. The Silicon Valley–Hsinchu relationship today consists of countless formal and informal collaborations between investors, entrepreneurs, small- and medium-sized companies, and the divisions of large companies located on both sides of the Pacific. Today the two regional economies are at different technological levels—Taiwan remains the world's most efficient and agile IT manufacturer, while the United States continues to define new standards, products, and technologies. These complementary specialties fuel a process of reciprocal regional upgrading.

Over time, Taiwan's political leaders came to view overseas Chinese engineers as an asset in their efforts to upgrade the island's position in the international economy. Taiwanese government officials began traveling to Silicon Valley in the 1960s, long before most of the world had heard of the region. They studied the Valley's experience and elicited advice on technology and policy from overseas Chinese engineers. By maintaining an ongoing dialogue with leading U.S. technologists during the 1970s and 1980s, economic policymakers in Taiwan laid the foundations for a development strategy that diverged fundamentally from policies being pursued during this time in most of East and Southeast Asia.[23]

The Chinese Institute of Engineers (CIE), the first technical association established by overseas Chinese based in the United States, provided an initial forum for collaboration between engineers on both sides of the Pacific. In 1966 Taiwan's Minister of Communications, Dr. Y. S. Sun, initiated the Modern Engineering and Technology Seminars (METS), held every two years and coordinated by the CIE/USA and its Taiwanese counterpart, CIE/ROC.[24] These week-long seminars were designed to bring leading engineers and scientists to Taipei to provide state-of-the-art expertise to local industry. Industry representatives in Taiwan would develop a list of topics of interest for the upcoming seminar, and CIE members in the United States would identify appropriate experts from industry or academia who would travel to speak on the selected topics.[25] In the early years METS introduced up-to-date technologies to Taiwan but also helped to create personal and professional relationships between the engineers based in the two countries. Over time, senior government officials associated with science and technology, many of whom were also U.S.-educated engineers, began attending the meetings as well. Finance Minister K. T. Li, considered by many to be the architect of Taiwan's technology strategy, was a regular METS attendee.

While policymakers initially viewed this overseas community as a source of technological expertise, they increasingly recognized its members' potential as policy advisers. There was, not surprisingly, an early consensus on the need to invest in education. The number of graduates of higher education in Taiwan grew from under 10,000 per year in 1961 to almost 200,000 in 1996 (constituting 18 percent of the population over age 25)—and 40 percent of the degrees were in engineering. This extraordinary proportion of engineering degrees reflected the strong emphasis on practical education among Taiwanese families (an emphasis found in other developing economies, including China and India). The expansion of higher education in the postwar decades was critical to the growth of Taiwan's IT industries. Successive generations of graduates of these programs were quickly absorbed by domestic industry.

Private and public institutions also developed specialized training courses and programs to meet the demand for skill and the rapid pace of technological change.[26] In addition, the Ministry of Education initiated a Masters-Degree-on-the-Job program to make higher education more accessible. The largest of these programs, run by the nonprofit Institute for Information Industry (III), trained more than 175,000 professionals between 1980 and 1996, primarily in six-month full-time classes.[27] Universities and government agencies also expanded opportunities for part-time training in highly specialized technical areas to accommodate the needs of employees. By the 1990s many of Taiwan's producers faced acute shortages of engineering talent and rising wages, and began providing stock ownership plans to attract and retain top workers.

This commitment to education and training distinguished Taiwan from its neighbors in Southeast Asia. When John Hsuan, the chairman of United Microelectronics Company (UMC), looked elsewhere for a location for a factory, for example, he concluded: "The number of technical people that Thailand educates and trains annually is not sufficient to meet the requirements of a wafer factory."[28] A limited supply of university-educated technical workers has also constrained IT growth in Hong Kong, Malaysia, the Philippines, and Indonesia. Acer veteran and venture capitalist Ken Tai expressed a common sentiment among Taiwan's industrial leaders when he said, "The best thing that the Taiwanese government has done in the past thirty years is its continued investments in education."[29]

Building Community: The Detailed History

As Minister of Economic Affairs, Dr. Y. S. Sun—who had initiated the METS seminars in the 1960s—worked with U.S.-based advisers to launch a domestic semiconductor industry. These conversations spurred the formation of the Industrial Technology and Research

Institute (ITRI) in 1973 as an autonomous agency dedicated to providing joint research, technical services, and advice to Taiwan's small- and medium-sized enterprises. The following year ITRI officials created the Electronics Research and Services Organization (ERSO), a subsidiary devoted to developing the domestic semiconductor industry, and in 1976 they recruited another U.S.-educated electrical engineer, Ding Hua-Hu, as the first director general.

Both organizations were initially government-funded, but Y. S. Sun—influenced by his U.S.-based advisers—insisted on increasing the share of private-sector participation in order to avoid endowing ERSO with a civil service mentality. By 1988 ITRI was receiving 55 percent of its funds from the public sector and the balance from the private sector, while ERSO received only 25 percent of its funds from the government, with the balance coming from fees from private companies for services. By 1987, ERSO had a staff of more than 1,700 and a budget of about $100 million in U.S. dollars.[30]

In 1974 Sun invited Dr. Wen-Yuan Pan, an RCA engineer active in METS and CIE/USA, to develop a plan for upgrading local industry, particularly the electronics industry. After meeting with industry leaders and top officials (including Premier Chiang Ching-Kuo and senior cabinet members), Dr. Pan recommended that Taiwan acquire foreign semiconductor technology. He established and led the Technical Advisory Committee, a strategic planning group of senior executives from leading corporations including Bell Labs and IBM as well as top university researchers in the United States, with a mandate to steer and oversee the development of Taiwan's IC industry.

Committee members met weekly in New York to develop recommendations and traveled to Taiwan regularly to work with personnel from ITRI and ERSO. In these meetings they provided information on industry trends as well as the experience of U.S. researchers with different approaches to IC development (including efforts that had proved to be dead ends). The advisers helped formulate the plan to transfer production technology from abroad as well as the decision to focus on manufacturing custom chips, or application-specific integrated circuits (ASICs), rather than capital-intensive large-scale semiconductor manufacturing. The identification of a niche market and of an emerging technology (CMOS, or complementary metal-oxide-silicon) allowed Taiwan to avoid direct competition with established industry leaders in the United States. This differentiated the Taiwanese from Korean and Japanese producers pursuing high-volume products like DRAMs and avoided worsening the economics of a semiconductor industry plagued by global overcapacity.

Initially, government and industry leaders in Taiwan objected strenuously to the plan to transfer IC technology from the United States. It was a very ambitious and costly project for the time, requiring an outlay of $12 million for the period 1975–1979. Acer's Stan Shih recollects:

> The advocacy for developing semiconductors in Taiwan started in the mid-1970s when several experts came back from the US to promote Large Scale Integration. At that time I did not agree with this. Taiwan, then, did not have enough capital and our capability in controlling the market was also insufficient. Moreover, Taiwan was not mature enough in terms of designing and manufacturing capabilities. I personally had no opinion if the government was to allocate budget to conduct research, but it would be too early for the private sector to make such a huge investment.[31]

The advocates of technology transfer prevailed, however, largely because of the stature and vision of the foreign advisers as well as the strong advocacy for the project by Y. S. Sun. Taiwan's National Science Council in turn began investing heavily in local IC research and training by establishing IC process laboratories and expanding training and research at local universities.

With the help of the U.S. advisers, ITRI selected the American consumer electronics company RCA as a partner for the transfer of semiconductor technology. In 1976, ERSO signed a five-year technology transfer agreement with RCA. The agreement stressed training as well as transfer of technology, so ERSO sent a team of thirty-seven engineers (recruited from both Taiwan and the United States) to an RCA facility in the United States for a year of intensive training in IC design and manufacturing. The agreement also stated that RCA would transfer all technological advances as well as design and process improvements for making CMOS devices, purchase back a certain quantity of wafers produced by the plant, and provide information on product applications.[32] This team (now known as the RCA-37) formed the core of the leadership of Taiwan's IC industry in subsequent decades.[33]

The group returned to Taiwan after a year of training to run ERSO's newly constructed IC pilot production facility. In 1977, they began designing and manufacturing chips on a small scale for commercial users in Taiwan. The plan was to transfer the technology out of the lab to the private sector by soliciting private investments, while also increasing production capacity and improving quality. However, it was difficult to find private investors because the IC industry was still viewed as extremely risky. In 1979 UMC became the first spin-off from ERSO; its technology, equipment, and engineers came directly from ERSO, while its financing came from both ERSO and five local companies. It was initially very difficult to find private investors in UMC, but the firm quickly became profitable. When it went public in 1985, the government was a minority shareholder (23 percent). Robert Tsao, one of the original RCA-37, became CEO of the firm. Over the next decade UMC became one of Taiwan's leading companies, and ERSO spun off several other semiconductor-related companies.

In 1979 Y. S. Sun became Premier of Taiwan. Not surprisingly, he sought to broaden and institutionalize support for high-tech projects, particularly semiconductors, through the creation of the Science and Technology Advisory Group (STAG), a small high-level group of independent policy advisers. With K. T. Li as its head, STAG's mission was to formulate action plans to accelerate the development of science and technology in Taiwan, reporting directly to the premier and his cabinet. K. T. Li recruited all fifteen STAG members from the United States: the group included prominent Chinese as well as non-Chinese engineers and executives from Bell Labs, IBM, and other large technology corporations, including some Technical Advisory Committee members.

STAG members returned to Taiwan annually for the National Development Conference and worked closely with their local counterparts in industry and government to learn about emerging problems and to create a policy consensus. The members of STAG became known in Taiwan as "foreign monks"—outsiders who gain special respect in Chinese society in their status as foreigners and as experts. The group was influential in directing resources toward engineering research education and training as well as continued upgrading of the domestic technological infrastructure. Their endorsement allowed the National Science Council to make major financial commitments to semiconductor research with local universities, despite strong opposition within the government.

STAG's aggressive and competitive approach to Taiwan's technology development was strongly resisted by the more conservative forces within the state. STAG argued, however, that if Taiwan did not develop state-of-the-art technology, its small- and medium-sized firms would be vulnerable to capture by large, established foreign producers. As with the initial decision to invest in IC capabilities, the opposition in the early 1980s to proposals to invest in Very Large Scale Integration (VLSI) technology and production was strenuous. The debate centered on the enormous costs (and forgone opportunities) associated with VLSI technology, as well as the substantial technical challenges associated with its development. (Japan had only recently started a VLSI program, and other countries had tried and failed.) In 1984, after a year-long debate, the project was approved and ERSO began investing in university VLSI design projects and collaborations.

The "foreign monks" also helped consolidate an industrial strategy that limited direct state intervention in favor of reliance on the private sector and market opportunities. Mathews describes this as "non-dependent development" that relies on "leverage and learning" rather than on either multinationals or established domestic firms.[34] STAG member Ta-lin Hsu, who earned a doctorate at U.C. Berkeley and worked at IBM Research Lab in the 1970s, reports: "We encouraged them to train people and then to let them go out and start their own companies."[35] This aggressive stance helped push Taiwan to develop private-sector manufacturing capabilities at an unprecedented pace—as evidenced most clearly in the 1985 formation of Taiwan Semiconductor Manufacturing Corporation (TSMC) as a joint venture between the government (49 percent), Philips (27 percent), and local investors (24 percent) to pursue the development of VLSI technology. The overseas Chinese community was crucial to the success of this venture. Morris Chang, the former CEO of Texas Instruments who had returned to Taiwan to run ITRI, played the lead role in convincing private investors, particularly Philips, and he recruited an executive from General Electric's chip division to run TSMC.

During this period the Taiwanese government offered a variety of grants, low-interest loans, and subsidies to foreign and domestic electronics firms. These incentives contributed to local industrial growth in part because they were universally available to all producers in a particular sector. At a time when European and East Asian governments were channeling resources to national champions, Taiwan's policymakers created a competitive environment by supporting multiple firms in any initiative, and constantly putting pressure on the established players in order to develop "natural" rather than "national" champions.[36] Although ITRI organized the initial IC technology transfer in the 1980s, it never mounted the very large scale collaborative industry/government research projects that characterized the Japanese and European Union industrial policy in the 1970s and 1980s.

ITRI established its first foreign office in Silicon Valley in the early 1980s; it was soon followed by other Taiwanese government agencies involved with science and technology policy, including the Hsinchu Science Park, the Science Division of the National Science Council, and the Institute for Information Industry. These agencies quickly built ties with local industry associations in order to monitor industrial and technological trends for domestic producers. They aggressively recruited overseas engineers to return to Taiwan by building databases of U.S.-based Chinese engineers and computer scientists and sharing them with talent scouts from Taiwanese firms, as well as by providing information and contacts to overseas Chinese considering setting up technology businesses in Taiwan. The National Youth Commission sponsored visits by overseas scholars and professionals and even financed the airfare for professionals who were returning to Taiwan permanently. When the Taiwanese government initiated major engineering projects—from a transit system to a power station—they consulted the U.S. branches of the CIE.

Reliance on U.S.-based Chinese engineers for technical and managerial expertise and policy advice fundamentally shaped the direction and pace of Taiwan's technology development. It also created close personal and professional relationships within a growing circle of engineers, entrepreneurs, executives, and bureaucrats on both sides of the Pacific—and contributed to the formation of shared worldviews and collective identities among Chinese engineers in both places. The list of recipients of the Chinese Institute of Engineers–

USA Annual Awards for Distinguished Service and for Achievement in Science and Engineering over the past three decades reads like a "Who's Who" of Taiwanese technologists based in the United States and Taiwan.[37] In short, an unintended consequence of Taiwan's outward-looking technology policies was the creation of a cross-regional technical community with a shared commitment to the economic development of Taiwan.

SUPPORTING ENTREPRENEURSHIP

Taiwan's highest-profile attempts to imitate the institutions of Silicon Valley were the creation of the Hsinchu Science Park (formally the Hsinchu Science-Based Industrial Park) and the development of a domestic venture capital industry. After observing the experience of U.S. technology regions, Taiwan's National Science Council proposed a science park as a means of attracting foreign investment in research-oriented companies. Despite limited interest from both private and public sectors, the Hsinchu Science Park was established in 1980 as a high-technology version of an export-processing zone, one that offered subsidized land and financial incentives for R&D-intensive manufacturing tenants. For example, government-owned banks offered to invest up to 49 percent of the shares in joint ventures and the Park Administration provided generous tax incentives to qualified (research-intensive) firms, including a five-year tax holiday followed by a maximum tax rate of 22 percent. Hsinchu City was chosen as the location because of its proximity to ITRI headquarters as well as to National Tsinghua University and National Chiao Tung University. The ERSO lab was moved to the area as well.

Despite aggressive recruitment by government officials, including a former Honeywell executive, the Science Park failed at its initial goal of attracting multinational investments. Ironically, American electronics companies were aggressively investing in assembly and other production facilities in Southeast Asia at this time; Taiwan was repeatedly bypassed in favor of Malaysia and Singapore. While it may not be surprising that the park's administrators failed to attract leading firms like Texas Instruments, they were not even able to attract Chinese-American firms like Wang Laboratories and Silicon Valley–based Wyse. The exception was AT&T, which established a factory in the park to assemble switching systems because this was a prerequisite for winning a government telecommunications contract.[38] After eight years, the Hsinchu Science Park remained a relative backwater—home to ninety-four companies, mainly domestic firms, including ERSO spin-offs like UMC and TSMC, whose collective sales totaled less than $2 billion.

Faced with the slow growth of the Science Park, government officials sought alternative development strategies. K. T. Li, who had visited Silicon Valley regularly in the 1970s to meet with Chinese engineers, was especially impressed by the U.S. venture capital industry and saw it as a potential missing link in Taiwan. In the early 1980s—long before it was fashionable elsewhere—he convinced the Ministry of Finance of the need to provide funding for research-intensive production and promote the development of a public capital market. Once again Taiwan's policymakers studied the U.S. experience: they consulted investment professionals, organized collaborations with large U.S. banks to share financial and managerial expertise, and sent teams to Silicon Valley to be trained in managing a venture capital firm. Clark Su, chair of the Taipei Venture Capital Association, describes venture capital in Taiwan as a "pure transfer" from Silicon Valley.

K. T. Li spearheaded the legislation to create and regulate the venture capital industry and set up a framework for enterprises to establish venture capital funds. This was no small task, since venture capital was completely foreign to traditional Taiwanese practices, dating back centuries, in which family members closely controlled all of a business's financial affairs. Under Li's guidance, the Ministry of Finance created significant tax incentives: 20 percent of the capital invested in strategic (technology-intensive) ventures by individual or corporate investors was tax-deductible. Recognizing the challenge of raising capital from Taiwan's risk-averse financial and industrial communities, the government also provided substantial matching funds. The Ministry of Finance organized the initial "seed fund" with NTD $800 million from the Executive Yuan Development Fund.[39] This was so quickly allocated that the government committed a second fund of double the original amount.

Acer founded Taiwan's first venture capital firm, Multiventure Investment, in 1984 as a joint enterprise with Continental Engineering Group. Equally important, K. T. Li invited senior Chinese-American financiers to establish venture capital companies in Taiwan. Ta-lin Hsu, who had been a key senior policy adviser and STAG member since the 1970s, set up Hambrecht & Quist Asia Pacific in 1986. Hsu reports that it was not easy to raise the initial $50 million fund. In particular, K. T. Li "twisted lots of arms" to raise $21 million (51 percent) from leading Taiwanese industrial groups such as Far East Textile, President Enterprises, and Mitac. The balance (49 percent) came from the government.[40]

The first general manager in H&Q Asia Pacific's Taipei office, Ding-Hua Hu, earned a doctorate in engineering at Princeton University in the 1970s; as the first general director of ERSO (and previously an associate director of ITRI and a professor at Chiao Tung University), he played a lead role in building Taiwan's semiconductor industry. His career—and connections—underscore the extent to which the social and professional networks cut across university, government, and private-sector (both financial and industrial) worlds in Taiwan and the United States. H&Q Asia Pacific's early investments included Acer, UMC, Microtek, and Tai Yan, and the successes of these investments made subsequent rounds of fundraising easier.

Two other U.S.-educated overseas Chinese engineers, Peter Liu and Lip-Bu Tan, responded to Li's invitation as well. They established Taiwan's second U.S.-style venture fund, the Walden International Investment Group (WIIG), as a branch of the San Francisco–based Walden Group in 1987. Both H&Q Asia Pacific and WIIG (along with Peter Liu's spin-off firm, WI Harper) remain leading investors in Taiwan's technology industries, and increasingly in mainland China as well. Once the early investments began to pay off, domestic IT firms created their own venture capital funds, including TSMC, UMC, Macronix, Winbond, Acer, UMAX, D-Link, Mosel, and SiliconWare. After that, even the old-line firms in traditional industries also began investing in IT-related businesses.

The availability of venture capital transformed the Science Park from its originally envisioned role as an export processing zone into an open laboratory for the growth of indigenous technology firms. This was a distinctive development in Asia at a time when capital was available only to large corporations with ties to governments or wealthy families. During the 1980s Taiwan remained a low-value-added producer of electronics components, but in the early 1990s, with the growth of entrepreneurship and accumulated production experience, local firms began to differentiate products on the basis of innovation and quality rather than simply cost.

THE BRAIN DRAIN REVERSES

The reversal of the brain drain—when thousands of U.S.-educated engineers returned to Taiwan to contribute to the development of its technology industries—was pivotal in transforming Taiwan into a high-skill, high-wage economy. Government agencies had aggressively recruited overseas Chinese engineers to return to Taiwan in the 1980s, but succeeded in attracting only a handful each year. The turning point came in 1987, when Acer went public in Taiwan. It was followed in 1988 by Microtek, a scanner company started by four returnees from California. The spectacular performance of these companies signaled to entrepreneurs that economic opportunities in Taiwan had the potential to be better than those available in the United States. Although the stock market boom coincided with the lifting of martial law in Taiwan, most returnees report that their decision to go back was primarily driven by the promise of economic opportunity.

Incentives for Return

The local stock market was especially attractive because Taiwan taxes employee stock on the basis of its price when issued rather than its current market value. In addition, Taiwan eliminated its capital gains tax in 1990, so no tax applied when shares were sold. Returning Taiwanese who brought capital gains from stock options or venture capital from another jurisdiction (such as the United States) were not taxed on those capital gains either.[41]

The technology recession of the early 1990s, which left many engineers in the United States unemployed, served as an additional motivating factor. In the period between 1986 and 1996, more than 40,000 overseas Chinese returned to Taiwan from the United States. Figure 4.2 illustrates the surge of returnees to Taiwan that started in the late 1980s and peaked between 1992 and 1995, when some 5,000 returned annually. The returnees quickly reconnected with former undergraduate classmates in Taiwan—many of whom occupied important positions in technology companies and agencies—while maintaining the ties they had established in the United States. Patrick Wang, who returned to Taiwan in 1983 to start a scanner

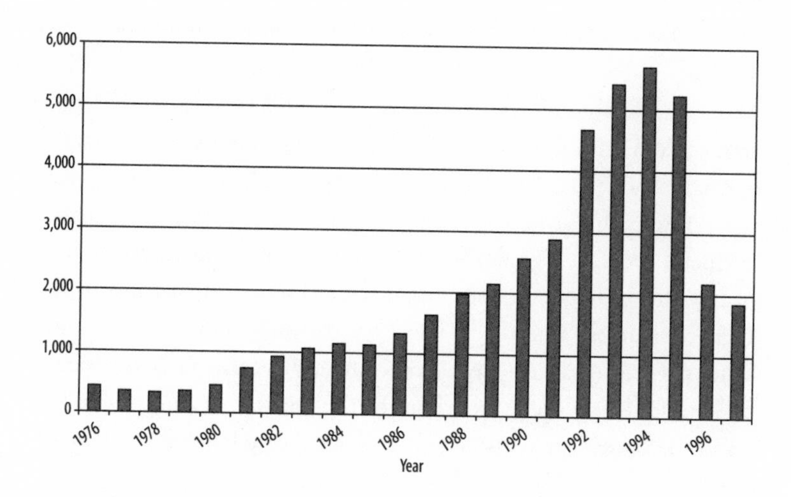

Figure 4.2. Returnees to Taiwan from the United States, 1976–1997.

company, emphasizes the importance of the "Stanford Mafia" to successful Taiwanese executives like himself who were educated at Stanford.[42] By 2000, there were 364 graduates of the Stanford Electrical Engineering Department working in Taiwan; there were also informal networks of ex-Bell Laboratories and ex-IBM employees who met regularly in Hsinchu.

These returnees to Taiwan, many of whom had worked for at least a decade in the United States, brought with them not only technical skill but also organizational and managerial know-how, intimate knowledge of leading-edge IT markets, and networks of contacts in the United States technology sector. In the words of Genda Hu: "The U.S. experience had a tremendous impact on my generation. We were exposed to world-class researchers and leading-edge technical opportunities. We observed firsthand the growth of the electronics and computer industries, and how business models and mentalities were changing. We also developed confidence in our knowledge and skill sets, and we developed the cultural know-how and language to communicate in both Taiwan and the U.S."[43] The returnees also brought a new ownership model to Taiwan, as described by UMC founder Robert Tsao: "The ownership and management models in the Hsinchu Science Park are revolutionary for Taiwan. The firms are not government- or family-owned; management owns [a significant part of] the firms and manages the firms. This is unique in Asia, and it has facilitated partnerships with U.S. companies; when U.S. firms try to work with companies in other parts of Asia they have great difficulty because of government or family ownership."[44] In addition, the returnees came to understand the value of transparency and performance-based incentives in business.

Today, returnees from the United States are overrepresented in senior management ranks in Taiwan. Morris Chang, who earned a Ph.D. at Stanford and worked for twenty years in the United States before returning to Taiwan, is one of sixteen senior executives at TSMC with American graduate degrees and work experience (out of a total of nineteen senior executives listed on the Web site). As Chang reported in 1997, "TSMC employs about sixty or seventy returnees from the U.S., and most are in very senior management positions. Their most important contribution is their experience in managing world-class companies like HP, Intel, and Bell Labs. They bring the disciplined management style of these businesses. Their ability to work in the U.S. and to speak English well is also critical because two-thirds of our customers are outside of Taiwan."[45] Although Robert Tsao, CEO of UMC, has no U.S. experience except as part of the RCA training, five of the top nine senior executives in the company hold American graduate degrees.

Returnees were also disproportionately likely to become entrepreneurs: for example, they were on the founding teams of 121 companies (42 percent of the total 289) that were located in the Hsinchu

Science Park in 2000.[46] These entrepreneurs often imported entire teams of engineers to help run their rapidly expanding operations. The financial incentives were substantial, particularly in the semiconductor industry: by the mid-1990s Taiwan salaries for senior managers were three or four times the U.S. level, and many firms offered tax-free stock bonuses. In the late 1990s TSMC engineers, for instance, earned $400,000 to $500,000 a year because of lucrative stock bonuses. Even janitors were earning $50,000 a year.[47]

Snowballing Prosperity

The returnee population in Hsinchu Science Park took off in the mid-1990s, along with the growing investments and company sales. By 2000 a total of 4,108 returnees from overseas worked in the Science Park—a number comparable to the 4,464 ITRI alumni working in the Park at the same time.[48] When asked about the Park, returnees report that preferential access to scarce, high-quality housing and the only Chinese-American school in Taiwan, both of which are located on Park grounds, facilitated the decision to return, but that professional opportunities were their primary motivation. In fact, in many cases the wives and children of returning executives remained in the United States rather than going back to Taiwan. The economic incentives to return only increased with time, as CEOs of local PC and IC companies became U.S.-dollar millionaires and billionaires. By 2000 the Hsinchu Science Park, now with more than a hundred thousand employees and output of $27 billion in U.S. dollars, was so crowded that the government began plans to build another park in the south.

The Taiwanese venture capital industry became self-sustaining in the 1990s as well, as newly wealthy entrepreneurs along with businesses from traditional industries (who had been reluctant to invest earlier) were attracted by the demonstrated possibility of high returns. In 1990 Taiwan was home to 20 venture capital firms. By 1999, the number had multiplied to 153. The total capital invested by these firms increased dramatically as well, reaching $43.5 billion in NT dollars ($1.3 billion in U.S. dollars) in more than eighteen hundred companies. Wen Ko, a leading Taiwanese venture capitalist and former Hewlett-Packard executive, sees similarities to the situation in Silicon Valley in earlier decades: "These rich engineers put seed money into high-tech prospects because they're young and willing to take risks. . . . They invest in what they know best. There's a high-tech gold rush feeling here. Very few countries have so many people willing to put money into high risk start-ups."[49]

Taiwan now has the world's third largest venture capital industry (only those in Israel and Silicon Valley are larger) and a dynamic domestic capital market. In 1997 some 30 percent of Taiwan's population owned stock, compared to only 10 percent in Japan. The trading volume of the Taiwan Stock Exchange was $1.3 trillion (the third largest after New York and London), and average return on equity for a listed electronics company was an impressive 26 percent.[50]

The infusion of managerial and entrepreneurial talent from the United States, together with the connections that returnees provided to foreign technology and markets, were decisive in the upgrading of Taiwan's manufacturing capabilities. Even Taiwanese who remained in the United States frequently chose to consult and moonlight on product development, providing market intelligence as well as links to U.S. customers and technology for firms in Taiwan. Domestic producers benefited from their growing role as manufacturing partners for some of the world's leading computer and systems producers. These OEM relationships, which originated with contracts from Silicon Valley–based overseas Chinese companies, provided the production information and know-how as well as the volume manufacturing experience that allowed Hsinchu firms to incrementally improve their capabilities. As the partnerships deepened over time and

Taiwanese firms took on growing responsibilities for design, they contributed to the upgrading of the regional industrial system.

The dynamic growth of Taiwan's technology sector in the 1990s eventually slowed the outflow of talent. Local career opportunities, as well as the expansion of local postgraduate programs, meant that fewer graduates of Taiwan universities chose the United States for graduate education. In the early 1990s more than 30 percent of the engineers who studied in the United States returned to Taiwan, compared to less than 10 percent in the 1960s and 1970s.[51] In 1999, for the first time, the number of Taiwanese earning science and engineering doctorates at home surpassed the number earned in the United States, reflecting both the fast-diminishing number of Taiwanese earning S&E doctorates in the United States (which peaked in 1995 at 1,300 and fell to 750 in 2000) and the simultaneous growth of S&E doctorates granted in Taiwan.[52]

ACER IN ACTION: USING COMMUNITY FOR COMPETITIVE ADVANTAGE

Acer is Taiwan's largest and best-known technology company, and its contributions to domestic growth are well recognized.[53] Less well known is the extent to which Acer's success rested, from the start, on close ties to the overseas Chinese community in Silicon Valley. Stan Shih and two classmates from National Chiao Tung University founded the company in 1976 (although it was called Multitech until 1987). They soon recruited Ken Tai, another Chiao Tung University classmate, who had also been working on microprocessors for Mitac. Their initial financing, like that of virtually all Taiwanese start-ups at the time, came from the personal savings of the founders (totaling $25,000 in U.S. dollars); their early business included consulting and serving as agents or distributors for the Silicon Valley–

based semiconductor companies AMD and Zilog. The company also became a distributor for an Indian company, Wipro, in the early 1980s.

The founders quickly realized that they needed an office in the United States to coordinate transactions and collect information about new products and technologies. In 1977 they recruited another Chiao Tung classmate, Edward Chang, from Hewlett-Packard in Silicon Valley and established a joint venture in which Chang had a 60 percent share. The partnership provided timely information and facilitated access to low-cost components. According to Stan Shih, who describes Acer's development as "a history of partnerships," this joint venture was a precursor to his "local touch" and "local majority shareholder" strategy. As he put it, "Local partners not only can share risks together but they are also familiar with the local market, and understand better how to reduce operational costs. More important, after the operation is localized, we will be able to attract more local talent to contribute to Acer." When they established new units in Taichung and Kaoshiung, for example, the firm contributed 40 percent and three local shareholders in each city jointly contributed 60 percent.[54] Co-founder George Huang recalls that it was impossible to lure Chinese engineers like Chang back to Taiwan during the 1970s, but they were an essential source of knowledge and contact for their friends and classmates in Taiwan.

As Multitech, Acer's early business was entirely trading and consulting. It was not lucrative, but consulting provided a substantial base of experience: in five years the firm designed forty microprocessor-based applications for customers in Taiwan and abroad. In 1980 Multitech engineers designed video terminals that were subsequently manufactured for export by ADI Electronics; this became the first microprocessor product exported from Taiwan in high volume, and also served as a model of a collaboration that leveraged the strengths and spread risks across the partners. Its income was based on royalties (sales volume), not costs plus profit. This reflected Shih's belief in the importance of working together:

> For a company to accomplish a long-term and sound development, mutual trust and [a] cooperative relationship must be established among the partners and employees. Taiwan's entrepreneurship is most often the result of collaboration among a group of friends or schoolmates who share the same ideals. In a structure like this, each partner has to specialize in certain areas, perform his job well, and build a consensus.[55]

Multitech began to develop its own microprocessor-based products by importing both components and development systems from the United States. In its early years the firm avoided computers and focused on other microprocessor applications because competitor Mitac was already producing both Intel computers and DEC minicomputers. Multitech launched its first computer product, the Microprofessor Learning Kit, in 1981 and succeeded in establishing a brand name in the export market.

The firm's timely introduction of successive personal computer models in the 1980s was possible only because of its partnership with a Silicon Valley company, Sun Technology (Suntek). Taiwan-born K. C. Shih (not related to Stan Shih) started Suntek in the early 1980s to develop one of the first UNIX-based workstations. K. C. Shih—who had studied at U.C. Berkeley and MIT in the 1960s before learning microprocessor design and development while working at DEC, National Semiconductor, and Zilog—had been unable to raise venture capital from Silicon Valley investors. However, Stan Shih recognized the value of the microprocessor technology and knowledge for Taiwan, and he invested in Suntek through the newly formed Multiventure Investment (which invested in Quasel at about the same time).

Transferring the technology to Taiwan was not easy even with this partnership. Multitech sent a team of six engineers to Suntek to transfer technology and to develop a 16-bit workstation, but the project was far too complex for them and ultimately failed. Two years later the same team returned to Silicon Valley, this time to develop a 32-bit computer, and in 1986 Multitech successfully introduced one of the world's first PCs built with the Intel 386 processor.[56] This represented a considerable achievement for Multitech, but Suntek went bankrupt (partly as a result of a lawsuit by National Semiconductor) and in 1987 K. C. Shih sold his workstation design to Acer. At around the same time Acer acquired San Jose–based Counterpoint Computer, a manufacturer of UNIX multiprocessor computers.[57]

Investments in firms like Quasel, Suntek, and Counterpoint not only provided access to leading-edge products and technologies but also expanded Acer's connections to the network of overseas Chinese engineers in Silicon Valley. These connections facilitated partnerships with leading chip firms, including Opti (a chipset design company founded in 1989 by a team of U.S.-educated Taiwanese engineers who had worked together at Chips & Technologies) and S3 (another 1989 C&T spin-off with an overseas Chinese engineering team that designed chipsets and graphics chips). These collaborations ensured Acer's early access to the latest chipset designs, from graphics and multimedia to audio and modem chips, which it integrated quickly into successive generations of PCs.

In the late 1980s Acer also established its own R&D center in Silicon Valley, with responsibility for developing prototypes and locating strategic partners. A decade later the lab's more than one hundred engineers had built a reputation for innovative research and a close working relationship with Acer Venture Capital that resulted in joint ventures with local firms. By 1990 most of Taiwan's leading PC firms—Acer, Mitac, Tatung, Advanced Data, and America Research Corporation—had R&D activities in Silicon Valley.[58] As the relationships between producers in Taiwan and Silicon Valley continued to deepen and diversify, Taiwan's technology leaders recognized that they were well positioned to exploit the emergence of open systems in computers. In Stan Shih's words:

> The standardization of hardware and software makes system integration relatively easy. As such the computer industry has entered a new IT age. One of the benefits it offers is that the risks are greatly reduced under the vertical disintegration structure. Furthermore under the disintegration mode every member is able to make continuous and effective breakthroughs in their respective fields of specialty, with the operating efficiency also increased. . . . The arrival of the new IT age brought increasing entrepreneurial opportunities of niche markets for computer manufacturers. On the other hand, there is also brutal competition in which companies are eliminated if they are not able to become the top players in their respective fields.[59]

Acer's Companions and Competitors

As Taiwan's producers increasingly saw their role as suppliers of specialized components, they recognized the continuing need for a presence in Silicon Valley, which remained their leading market at the end of the 1980s. In the subsequent decade Taiwanese firms developed partnerships both locally and globally in order to monitor and respond quickly to the rapid pace of technical change in the computer industry.

Taiwan's first PC firms emerged out of the local video game industry—scores of small companies manufactured designs copied from Japan until 1982, when the government banned the playing of video games for "moral reasons." The shift from making video game machines to PCs was relatively easy, as were simple microprocessor-based products. By 1983 there were more than a hundred small firms in Taiwan, including Multitech and Mitac, that were cloning early Apple and IBM PC systems. While Apple ultimately clamped down on pirated versions of the Apple II, the openness of the IBM architecture combined with lax IP protection by the Taiwanese government facilitated the proliferation of small- and medium-sized PC companies in the Taipei area. The only part of the IBM PC that was not widely available off the shelf was the ROM-BIOS (which links and coordinates all the hardware components with the operating system and applications software). Taiwan's producers had not yet developed the technological capability to design the BIOS, so they acquired it either through illegal imitation or, after 1985, from specialized U.S. firms that reverse-engineered it legitimately.

The overseas Chinese community in Silicon Valley provided Taiwan's first OEM contracts for PCs. David Lee, the Chinese-American founder of Qume, a producer of daisy wheel printers, went to Taiwan in 1982 in search of low-cost manufacturing labor and a financial package that included attractive loans and tax benefits. He placed orders with three small Taiwanese companies, Multitech, Mitac, and Compeq, and sent a team from Qume to teach Taiwanese engineers how to manufacture and test IBM PCs.[60] These products were quite simple and required minimal production and design skill. Multitech, which was still operating out of a small apartment at the time, leased an old garment factory in order to fulfill the Qume contract, and Mitac could only accept a contract to make $15 million worth of PCs even though Qume wanted to order twice as much.

The success of these early contracts led a widening range of U.S. customers to Taiwan for the manufacture of everything from disks, printers, terminals, monitors, peripherals, and other related parts to complete PCs. These initial OEM orders were motivated by the substantially lower cost of skill in Taiwan—in 1985 unskilled production workers' average hourly wages were about 15 percent of those in the United States: $1.36 compared to $8.37.[61] These new enterprises relied on the supplier base that had been developed in Taiwan through earlier foreign investments in the production of calculators, pocket radios, televisions, printed circuit boards, and other consumer electronics. When multinationals like Ampex, Sprague, TI, and RCA closed their operations in Taiwan because of rising wages in the late 1980s, they freed up skill and know-how that further strengthened the capabilities of local firms.[62]

Over time these small Taiwanese producers would identify and enter lucrative niches in the market, and—faced with swarms of new entrants in low-barrier-to-entry businesses—were forced to steadily upgrade their capabilities. Although know-how and technology were transferred initially through OEM contracts, an informal process of technological learning-by-doing came to dominate; engineers first gained familiarity with the manufacturing process and then experimented by imitating and modifying circuit designs, and by making incremental improvements in products and in the debugging processes necessary for volume manufacturing.[63] In this way Taiwan's domestic firms eventually established a position as low-cost mass manufacturers. The flow of new competitors also ensured ongoing experimentation and a deepening division of labor that contributed further to collective learning and upgrading.

The Biggest Incubator of All

Acer contributed to the decentralization of the local PC infrastructure by supporting spin-offs and internal start-ups. Taiwan scholars describe Acer as a "mega-incubator" because it has provided capital, technology, human resources, and management skills to so many spin-offs.[64] Venture capitalist Ken Tai compares it to Silicon Valley's Fairchild in its proliferation of local offspring. Most of Taiwan's motherboard companies, for example, have Acer lineage. Asustek

Computer, which was started in 1989 by four engineers from Acer's R&D lab with financing from its Multitech Investment Fund, is a model of the sophisticated specialist companies in Taiwan's PC industry, winning top international awards for its motherboards and graphics cards. In 2002 Asustek received four awards from Tom's Hardware Guide, the leading independent Web site tracking the global hardware industry, including Best Overall Motherboard, Best Motherboard Maker, Best Mainstream Graphics Card, and Best Graphics Card Maker. Asustek continually refines its designs and production system in order to remain a faster, higher-quality, and more efficient high-volume manufacturer than its competitors. The firm also invests in R&D in order to add value to the core technologies for its major products, rather than simply expanding sales or market share.[65]

Acer's permissiveness about employee mobility has contributed to interfirm and interindustry social networks that are reminiscent of those in Silicon Valley's open labor markets. Former Acer employees occupy senior positions in many of Taiwan's PC and PC-related businesses, and the general attitude in the industry, as expressed by a UMC general manager, is that "the mutual flow of talent is normal."[66] The firm's business practices spread with its network of employees. Acer established an employee stock purchasing program in its early years ("to create an environment in which the employees prosper together with the company," says Shih), and there was a strong commitment to sound financial management and rejection of the traditional family-run business model. Shih recalls that "my most important mission after the establishment of Multitech . . . [was] how to maintain the advantages of a family business and eliminate its shortcomings at the same time."[67]

Acer also created a unique network of social contacts within Taiwan's PC industry through its early training programs. Beginning in 1978, Multitech instituted training courses for local computer engineers in order to develop a market for the firm's products. In Stan Shih's words:

> It was quite an effort to promote such products, as the general public had no idea what a microprocessor machine was . . . some customers presumed we were selling some kind of machines for gardening when we paid them a visit. Some salespersons even called on us to sell gardening books. . . . We dubbed ourselves the "gardeners of microprocessor machines" to promote the new concept and cultivate the potential market.[68]

Multitech established training centers in Taiwan's three largest cities to teach engineers how to use assembly language to control microprocessor-based circuits (for applications such as machine control of traffic light signals). The firm also started a monthly publication called "Gardener's Words" to educate consumers about the technology and applications of microprocessor-based machines. As Shih puts it, "these strategies require long-term efforts but not large expenditures," and they had a profound impact on the industry. In the first four years of the training courses, some 3,000 engineers completed the fifty-hour course. In a short time there was demand for the classes from large government and private-sector organizations as well, and the local market gradually developed. The monthly "Gardener's Words," with a circulation of 20,000 at its peak, in turn allowed the firm to maintain contact with customers. Similar projects continued in the 1980s: in 1986 the firm sponsored a "one thousand computer classroom" activity that attracted a hundred thousand people to try a hands-on experience with a PC. The training centers created an extended network of engineers able to use computers, and many of the firms' lecturers became important executives at Acer and were able to draw on a large customer base of former or current students who shared the same language.

Acer managers report that they are far more likely to contact former teachers, classmates, or colleagues when they encounter a business problem (related to engineering, marketing, or manufacturing) than to turn to a private consulting firm or institution.[69] According to D-Link vice president Eric Wang:

> Most high-level people in IT in Taiwan are very good friends. Most are high school or university classmates, and this is a very special relationship. From the purview of know-how it means we share our ideas with one another. We don't have the resources to develop all of our own ideas, so we can share manpower or ask for investments from each other. For example, we might want to do a project for the small home office market such as an ISDN router. We lack the ISDN knowledge that the other company has, and they lack the LAN technology that we have, so we jointly develop the product. This is common with Taiwan partners.[70]

Of course shared experiences, particularly in a highly localized community, can also intensify competition as well. A top executive at TSMC reports that given their shared background, "There is no way NOT to expect competitiveness from such work teams!"[71]

5 ❖ Taiwan as Partner and Parent

Taiwan's success as a technology imitator made it an increasingly important partner and collaborator for Silicon Valley. Moreover, unlike Japan in the 1980s, Taiwan followed imitation with differentiation, not with direct competition. By concentrating initially on the manufacture of ASICs and specialized components for personal computers, Hsinchu built capabilities that were complementary to those in Silicon Valley, where shortages of space and skill made manufacturing increasingly uncompetitive.

A community of U.S.-educated Taiwanese engineers pioneered cross-regional collaborations during the 1990s. Winbond's Fred Cheng, for example, had worked in Silicon Valley for twenty years but knew Taiwan's technology community as well because he traveled to headquarters at least ten times each year. Conversely, Mitac vice president C. S. Ho never studied or worked in the United States but claims to know the Silicon Valley Chinese networks better than most people who work there. He reports: "There were a hundred of us in my graduating class at National Taiwan University. Only fifteen of us remained in Taiwan, and all but two of those who left went to the U.S. When I started traveling to the U.S. regularly I realized that there were about forty members of my graduating class still living in Silicon Valley. . . . Last week in Silicon Valley, I met with thirty different companies in a single week." As an investor, Ho depends on timely information: "My people network provides early warning; once news is in the newspaper it is too late."[1]

Ken Tai, one of the first employees of Multitech, is a good example of the new breed of investors who facilitate cross-regional collaboration. Tai worked for seventeen years at Acer and then for two start-ups before starting his own venture capital firm, InveStar, in 1996. In its first year of operations, the firm invested $50 million in Silicon Valley companies. Tai describes InveStar's role as an intermediary linking Silicon Valley and Taiwan: "When we invest in Silicon Valley start-ups we are also helping to bring them to Taiwan. It is relationship-building . . . we help them get high-level introductions to the semiconductor foundry and we help establish strategic opportunities and relationships in the PC sector as well. This is more than simply vendor-customer relationships. We smooth the relationships."[2]

SEEDING AN INDUSTRIAL ECOSYSTEM

As we have seen, Taiwan's technology industry grew and diversified in the 1980s and 1990s largely through collaboration between local producers and those located in Silicon Valley. The cases of Microtek and D-Link illustrate this process of specialized infrastructure creation in Taiwan—one in the scanner industry and one in networking.

Taiwan imported its scanner industry from the United States. The founders of Microelectronics Technology (Microtek) were U.S.-educated Chinese engineers who worked together in the 1970s on document engineering, laser printers, and scanners at Xerox. They were also among Taiwan's earliest returning entrepreneurs: they established Microtek International in Hsinchu in 1980 with a liaison office, Microtek Lab, in California. Microtek's first products were in-circuit microprocessor emulators, which sold well and won a prize for the best product at Wescom (a computer industry trade show) in 1981. The founders soon recruited additional returnees from Xerox and by 1983 had broadened the product line to include scanners. Microtek went on to produce the world's first 300-dpi black-and-white sheet-fed scanner in 1985 and remained an industry leader throughout the 1990s by introducing affordable scanners using innovative software and design.

One of Microtek's biggest initial challenges was the absence of capable suppliers for manufacturing scanners in Taiwan—an especially pressing problem because their main competitors were Japanese firms that had access to a sophisticated domestic infrastructure. Founder Robert Hsieh and his fellow organizers decided to nurture the development of local subcontractors and suppliers. They located a local company that was making low-end lenses and helped it upgrade its capabilities, first by developing a Microtek lens design and then transferring the technology to the firm (one Microtek founder holds fifteen patents in lens design). Over time, they helped to create an independent, world-class lens manufacturer. Similarly, there was no high-precision machining available in Taiwan in the early 1980s, so Microtek's founders worked closely with local machine shops to gradually upgrade their capabilities. Hsieh reports that they couldn't initially hold them to U.S. standards, but they worked out ways to jointly develop products with lower tolerances that still achieved high performance levels. As a result, Microtek was the first company to be awarded ISO 9001 certification for its scanners.

By 1988, when Microtek went public in one of the first technology IPOs in Taiwan (soon after Acer), the firm had helped to create a sophisticated domestic scanner infrastructure. The only components that were not locally available were charge-coupled devices, but the large numbers of competitors in the sector ensured continued supply for Taiwanese producers. In 1989 Microtek invested in Ulead Systems, a Taipei-based image processing and digital video software development start-up that was co-founded by three colleagues from Taiwan's Institute for Information Industry (two were also classmates from Chiao Tung University). Robert Hsieh describes how Microtek helped Ulead in its early days by jointly purchasing charge-coupled devices from Kodak in order to allow Ulead (which needed only small volumes) to take advantage of its substantial volume discount ($150 per device as compared to $500). This partnership benefited both firms and contributed to further upgrading the component infrastructure in Taiwan. In 1999 Ulead became the first publicly traded software company in Taiwan. It now has 450 employees and sells image editing, video editing, and 3D products worldwide.

In 1995, Taiwan-based producers accounted for 60 percent of the world scanner market; by 2000 the share had grown to almost 90 percent. Microtek alone accounts for over 20 percent of the world scanner market, and it has diversified into related digital image processing products such as digital cameras, LCD flat-panel monitors, LCD projectors, and PDAs. The firm's commitment to advancing technology is evident in its two R&D labs (one in California and one in Taiwan), which have facilities dedicated to optics design, mechanical and electronic engineering, software development, and product quality. The days when Microtek's competitive edge depended on low labor costs are long gone. Today the firm holds more than a hundred patents worldwide and continues to dedicate more than 20 percent of staff and 10 percent of revenues to R&D on next-generation digital imaging solutions.

In addition to nurturing relationships with innovative partners in Taiwan, Microtek has maintained extensive connections to the United States. The firm established an advanced research laboratory in San Jose in 1995 to support the work of a very talented Taiwanese engineer. This thirty-person team works closely with large Silicon Valley customers to develop promising new technologies and products. Microtek also set up a unit in Hillsborough, Oregon, to be close enough to Intel's microprocessor facility to co-develop emulators.

D-Link Corporation, one of Taiwan's earliest and most successful data network equipment companies, was started in 1986 by seven graduates of National Chiao Tung University. Its first product was a 1Mbps network interface card (NIC). The firm set up its first factory in Hsinchu Science Park in 1991, and was listed on the Taiwan Stock Exchange in 1994. Today D-Link is the world leader in wireless and ethernet networking for both consumer and small office, home office (SOHO) markets. The firm competes intensely with Linksys, which was acquired by Cisco in 2003, and with NetGear. Each maintains a network of Taiwan-based suppliers.

Vice president Eric Wang, who earned a master's degree in the United States and worked in Silicon Valley before joining D-Link, relates that soon after its formation, the firm invested in two IC design houses in Taiwan, Myson and Tamarek, both of which have become important technology suppliers. D-Link also invested in two Hsinchu component manufacturing companies: Abocom, a PC card and fax modem producer, and Cameo Communications, a high-end maker of LAN products. According to Wang, the relationships between D-Link and these suppliers are extremely close: "Abocom or Cameo could easily be part of D-Link. Each one is an independent company, but our share is about 40 percent or so and we have very close links with them. This is quite typical in Taiwan. Acer is similar with its 'cellular' architecture."[3] Another key partnership is with the semiconductor foundry UMC. Robert Tsao, CEO of UMC, is a very close friend and classmate of the D-Link president.

Like most Taiwanese companies, D-Link has an R&D center in California. This ensures access to components for very new technologies that are not yet available in Taiwan so that the firm can design them quickly into prototypes. D-Link also works closely with its U.S. networking customers. These collaborations are made significantly easier, according to Wang, because so many big Silicon Valley companies have Taiwanese engineers doing hardware design. He also describes the case of a Taiwanese chip designer who left the U.S. networking firm Synoptics and worked for D-Link for two years developing new products, and then returned to Silicon Valley and established his own start-up. D-Link invested in the firm, which develops products and licenses the technology to D-Link and other Taiwanese networking companies. D-Link also has a venture capital arm that invests in Bay Area companies in the networking field. Wang says that he travels to the United States six times a year and claims to know Silicon Valley well enough to drive without a map. The ties to Silicon Valley are extremely important sources of new technology in an industry like networking that is changing so rapidly.[4]

In 2003 D-Link spun off its OEM/ODM business to create a new contract manufacturing company named Alpha Networks. Alpha owns D-Link's production sites in Taiwan and southern China as well as its Ethernet technology, and remains a major supplier to D-Link. D-Link will focus on new products under its own brand. In 2004 D-Link sold 17 percent of the shares in Alpha Networks to Quanta Computer, signaling an expansion of their partnership. The chairman of Alpha Networks explains: "Our cooperation with Quanta Computer, be it products, manufacturing, cost containment and global logistics, [is] extremely complementary in light of the convergence of the computer and network communication industries."[5] Quanta benefits from Alpha's networking and high-speed Internet knowledge; Alpha gains access to Quanta's computer system integration and manufacturing capabilities.[6]

UPGRADING THE SYSTEM

The social networks in Taiwan's technical community are crucial to coordinating business relations in its complex and fragmented production supply chain. The production of a laptop computer, for example, typically involves at least a hundred different suppliers. Specialized producers compete intensely and may also collaborate, at least for the length of a given project. The close but shifting collaborations among networks of suppliers and customers are essential when time-to-market and time-to-volume determine competitive success, and when specialized components are continually upgraded and recombined into new end-products.

PC Partnerships

Continuous capability building allowed Taiwan to maintain its position as the dominant manufacturing base in the global PC industry during the 1990s in spite of rapidly rising labor costs.[7] Local industry leaders also point out that Taiwan contributed to the victory of the IBM PC over Apple. By reducing costs and facilitating speedy migration from one generation to the next, the Taiwanese producers accelerated the growth of the PC market. Ken Tai—now a venture capitalist—explains:

> When I was at Acer we approached Apple about manufacturing the Mac. They were producing in very large volumes at the time, but they had a top-down and inflexible model in which volume negotiations were based on a one-time contract. This didn't leave us any opportunity to innovate and raise our prices. The PC model was bottom-up: by allowing competition among many vendors, they encouraged innovative improvements in quality and performance. The PC industry's strengths are not simply Intel and Microsoft, but also the continuous upgrading by Taiwanese producers that eventually were able to provide Mac-level performance at substantially lower prices.[8]

The flexibility of Taiwan's industrial system—with its continual entry and exit, competition, and incremental upgrading—also helps to explain how Taiwan weathered the Asian financial crisis that began in 1997 far better than its Asian counterparts. In contrast to other Asian economies, Taiwan's legal system makes bankruptcy relatively easy, and entrepreneurs are accustomed to starting again after their ventures collapse: "[They] tend to cluster in less capital-intensive industries and to subcontract the production of many of their components. That means they do not need to spend much on fixed assets such as machinery, which, in any case, can easily be sold off to rivals if a firm goes out of business."[9]

Taiwan reached the technical forefront in PC manufacturing in the early 1990s. By taking on a growing volume of OEM contracts, local firms mastered the speedy design of motherboards and the responsive and efficient integration of entire PC systems.

The dramatic increase in ISO 9000–certified companies in Taiwan, from none in 1990 to almost sixteen hundred by 1999, signals the growing focus on quality and production control in addition to flexibility and low cost.[10] Even as the pace of technical change increased and profit margins diminished, forcing some local producers out of business and others to go to China for lower-cost labor, local firms continued to shift into higher-value activities. The advent of the under-$1000 PC reflected the intensity of competition in an increasingly commodity-like market.

Taiwan's lead firms maintained their competitive advantage by moving beyond arm's-length OEM relationships to take greater responsibility for new product design and development. In the new original design manufacturing (ODM) relationships, the manufacturing partners controlled the entire supply chain for a product—from design, engineering, and production (including component procurement) to inventory management and logistics. This not only contributed to Taiwan's growing capabilities in logistics and supply chain management but also significantly expanded the opportunities for Taiwanese firms to interact with, and learn from, their customers.

By focusing on production, engineers in Taiwan developed deep manufacturing process, design, and system integration capabilities, while U.S. engineers continued to pursue leading-edge technologies and features. Local electronic systems and PC companies increased the pace of new product introduction, improved yield and quality levels, and implemented scale-intensive mass manufacturing systems. By 2001 Taiwan, a country with a population of 22 million people, was producing 27 million desktop PCs per year and dominating the global market for notebook computers.

A detailed investigation of Taiwan's PC cluster in the late 1990s concluded that as foreign buyers demanded ongoing improvements in quality and production control, they became a critical source of knowledge for new product design and development. Demanding customers helped Taiwanese companies deepen partnerships with local suppliers and strengthen informal mechanisms for learning and diffusing knowledge within the cluster.[11] These collaborations are two-way: the brand-name holder might provide a road map, such as high-level product and performance specifications (screen size, microprocessor speed, and so on), or the Taiwan partner might generate a new design or product proposal for the customer. As a result, Taiwan's leading PC vendors frequently do more than 50 percent of the design work for name-brand computer companies in the United States. The partners in this ongoing joint process can develop a new model in six to nine months.

John Paul Ho, founder and general manager of Crimson Investment, a Silicon Valley private equity firm, credits Dell Computer's success in the PC industry to its ongoing collaborations with Taiwanese ODMs. Dell worked closely with suppliers like Quanta, Compal, and Lite-On, sharing information on technology and markets through regular meetings throughout the product life cycle. These interactions facilitated the ongoing integration of Dell's expertise with that of its suppliers to quickly build new capabilities, and allowed Dell to surpass more established producers in the United States, including IBM, HP, and Compaq, as well as its Japanese competitors.[12]

Although Taiwan's producers were technological imitators, their operational and engineering excellence as well as speedy integration and implementation of new designs allowed them to dominate markets for a growing range of PC and electronic systems. These firms built on accumulated experience in high-end precision manufacturing and close relationships with their customers and local component makers (including IC designers and fabricators) to quickly design and efficiently manufacture very high-quality products at lower cost. Acer senior vice president George Huang points out: "Nobody controls the pace of new product release in the PC industry; firms must be very fast to follow new trends and developments. A new component, software, or design can emerge from anywhere and late adopters will lose out. Momentum and flexibility are essential in this business."[13] And Gene Sheu, executive vice president of the Taiwan manufacturer First International Computer (FIC), comments, "Effective alliances and supply chain management and logistics are interdependent, particularly with the low-cost PC."[14]

A Compaq executive describes these partnerships from a customer's perspective:

> In the early years of this decade [the 1990s], Taiwan played a critical role in Compaq's reinvention. Compaq was born in the 1980s as a cloner of IBM PCs. Compaq enjoyed great success as the manufacturer of the "better" PC, the PC with bells and whistles, selling at a premium over other IBM compatible PCs. In the early 1990s consumers were no longer willing to pay a higher price for incremental features. Enter Eckhard Pfeiffer, new Compaq CEO. Mr Pfeiffer made the critical, strategic decision of turning to Taiwanese

> suppliers, often as sole source suppliers. That trust was paid back many times, with an extraordinary record of quality, reliability, and price competitiveness. In partnership with our Taiwanese suppliers, Compaq became the number one PC vendor in the world. Today Compaq is spending well over $7B dollars with 40 Taiwanese suppliers across our product range, and involving as many as 5,000 Taiwanese sub-suppliers in our business. This is an extraordinarily successful partnership.[15]

In 1999 Compaq launched an Internet-based service network to increase interaction between its suppliers and OEM partners in Taiwan and to control inventory levels. By 2000 more than twenty major Taiwanese manufacturers were participating in the Compaq Service Network.

Since 2000 U.S. computer makers have also used their partnerships with Taiwanese firms to enter new markets. Hewlett-Packard, which purchased some $16.2 billion worth of IT products from Taiwan in 2003, has ODM relationships not only in notebook computers, desktops, monitors, printers, and servers but also in flat-panel displays, PDAs and PDA handsets, MP3 players, and digital cameras.[16] HP's collaboration with Taiwan's Tekom in digital cameras, for example, has helped push the latter's capabilities in the design and manufacturing of higher-performance cameras, while strengthening HP's competitive position relative to the leading Japanese camera companies.[17]

Commanding IC Manufacturing

By the mid-1990s Taiwan's leading IC foundries, TSMC and UMC, had achieved technological parity with the United States and Japan, while remaining the world's lowest-cost IC manufacturers. The concentration of IC manufacturing expertise in Hsinchu was sufficient to lure IC designers to Taiwan. K. Y. Han, co-founder of the Silicon Valley–based Integrated Silicon Solution Inc. (ISSI), reports:

> We moved manufacturing to Taiwan because of logistics. We interact daily with our vendors, and it's much easier now because it's only a ten-minute drive to the fab and an hour to the assembly facility. Proximity wouldn't be so important if we were using older-generation, mature technologies, but our new products use leading-edge technology that requires continuous and intensive problem solving by teams of engineers to ensure the highest possible quality and performance. Our product cycles are down to six to nine months so we have to work well together with our partners.
>
> TSMC is a very close partner. We know them inside out and outside in. This is not just a customer-vendor relationship. It is a process of joint product and process development that requires a very deep relationship. TSMC discusses with us what next-generation products and technology we should jointly develop, so for example if we believe the next generation will include low-power requirements, communication capabilities, and must be portable, then we'll work with them to define the best process. Our products are advanced so they come to us to learn how the markets are evolving; we depend on them for high-performance and high-quality implementation.[18]

Cross-regional collaboration between fabless IC design firms and their foundries has deepened significantly in recent years. The partnership between Xilinx, a Silicon Valley designer of programmable logic devices, and Taiwan's UMC began in 1995. Ten years later the firm had 100 engineers on site at UMC, working with their engineering teams to develop new products. Xilinx attributes its sustained ability to "push the envelope" technologically to its partnerships. The firm was the first company to ship a device based on 90-

nanometer (NM) process geometries, and it is the leading producer of field programmable gate arrays (FPGA), with over 50 percent of the market.[19] Programmable devices require the smallest geometries and most advanced processes of all devices, and producers are under pressure to continue improving power management and security in addition to the perpetual cost and time-to-market pressures.

B. C. Ooi, Xilinx vice president for operations, emphasizes: "Our partners must share our view on risk taking. They must be willing to engage in joint risk taking, and be willing to try new things. It is the only way to stay on top of leading-edge technology, to truly innovate." Vincent Tong, vice president of product technology, adds that a Xilinx partner must be a "technology leader." He explains: "We can't afford to partner with companies that aren't leaders in what they are doing. Three or four years ago, when it came to manufacturing, you could throw a product design over the wall. Today, we need to be engaged tightly with our partners to drive technology forward. We have more than 80 process engineers, and we don't even have a fab."[20]

Taiwan's IC foundries also collaborate closely with their equipment suppliers in Silicon Valley. Steve Tso, a senior vice president at TSMC, worked at equipment vendor Applied Materials in Silicon Valley for many years before returning to Taiwan. He claims that his close personal ties with senior executives at Applied Materials provide TSMC with a distinct competitive advantage by improving the quality of communication between the technical teams at the two firms—in spite of the distance separating them. The interactions between TSMC and Applied engineers are continual, according to Tso, and, for the most part, are face-to-face because the leading-edge processes are not standardized and many of the manufacturing problems aren't clearly defined.

Tso travels to Silicon Valley several times a year, and reports that teams of TSMC engineers can always be found in the Applied

Materials' SV facilities for training on the latest generations of manufacturing equipment. Likewise, engineers from Applied regularly visit TSMC. Tso argues that this close and ongoing exchange helps TSMC develop new process technologies quickly while minimizing the technical problems that invariably arise when new manufacturing processes are introduced. It also keeps his firm abreast of the latest trends and functions in equipment design.[21]

David Wang, executive vice president at Applied Materials, summarizes the firm's model of partnerships with its customers:

> A customer's success is our priority. . . . If a customer starts to build a 300mm wafer facility, we also hire people to jointly develop the new facility. Joint efforts provide two advantages to our customers. One is a return asset. To earn profits as soon as possible, a rapid ramp-up is essential. As a partnership, we have to work at training our customers to operate the installed equipment, and at supporting process management, yield management, and in-situ methodology.
>
> The other advantage is short time-to-market. If a customer develops a new cell phone chip to be shipped within the next six months, we will work together with the customer to help with their process integration. A copper damascene process, for example, has complicated steps, such as barrier metal deposition, seed copper layer deposition, electroplating, polishing, low-k material deposition, and diagnostics. If a customer were to attempt to establish such major process integration alone, it could take them as long as six years. Working in a partnership, however, the customer has helped with the equipment installation and spare parts management.
>
> Ten years ago, a customer would initially have wanted to talk about a discount on the price of equipment. Now, the first thing they want to discuss is how we can work together.[22]

The adaptive capacity of Taiwan's technology base derives from collaborations among local producers as well as from their long-distance partnerships. The fragmentation and localization of production in the Hsinchu region are key to the flexibility, speed, and innovative learning-by-doing of its IC as well as its PC firms. In a report to investors, one company described the advantage of being part of the local cluster:

> The growth of the semiconductor industry has been the result of several factors. First, semiconductor manufacturing companies in Taiwan typically focus on one or two stages of the manufacturing process. As a result these companies tend to be more efficient and are better able to achieve economies of scale and achieve higher capacity utilization rates. *Second, semiconductor manufacturing companies in Taiwan that provide the major stages of the manufacturing process are located close to each other and typically enjoy close working relationships. This close network is attractive to customers who wish to outsource several stages of the semiconductor manufacturing process. . . .* Third, Taiwan also has an educated labor pool and a large number of engineers suitable for sophisticated manufacturing industries such as semiconductors.[23]

Taiwan's concentration of sophisticated IC specialists, as shown in Figure 5.1, looked to an industry observer from Japan like a single large enterprise: "In Taiwan [in 2000 there were] 130 design companies, 5 mask companies, 20 wafer manufacturing companies, and 100 assembly and test companies. Those companies work together and as a whole to form what is like one vertically integrated company. We can say that Taiwan is the world's largest semiconductor production base that resembles Silicon Valley of the US."[24] This observation, however, overlooks precisely those features that make Taiwan's industrial system so much more dynamic and flexible than the integrated Japanese corporations: the intense competition among

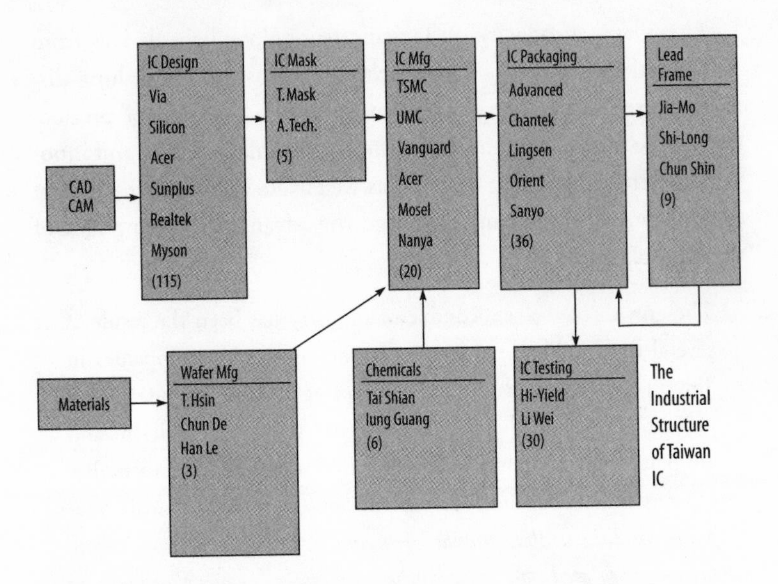

Figure 5.1 The structure of Taiwan's IC industry. Numbers in parentheses are the number of firms in each area as of 1999. *Source:* Electronics Research & Service Organization / Industrial Technology Research Institute, Hsin-Chu, Taiwan, 1999.

the producers in each specialized area of IC production, as well as their ability to spread risk and learn through shifting collaborations with local as well as long-distance customers and suppliers.

As the Hsinchu-based IC manufacturing industry expanded and matured, it became increasingly integrated with the agglomeration of PC and component makers in the Taipei metropolitan area. In the 1990s chip designers began to collaborate more closely with PC and peripherals manufacturers to design application-specific chips for their products, and these specialized chips were in turn manufactured at local foundries. By 2000 Taiwan was the largest and most sophisticated center of specialized IC design outside of Silicon Valley, with more than 130 companies including highly profitable firms like VIA Technologies, SiS, Realtek, Sunplus, and MediaTek. (The independent IC design industry in Japan and Korea, by contrast, was very small and underdeveloped.) Taiwanese IC packaging and assembly firms, such as Advanced Semiconductor Engineering and Siliconware Precision Industries, likewise became market leaders and accounted for over 30 percent of the world market.

Entrepreneurs who defined new products relied on past partners as well as new suppliers and designers to develop and manufacture these items quickly. According to Samuel Liu, Stanford-educated returnee and co-founder of Silicon Integrated Systems (SIS), a leading IC firm, "We are only able to be first to market because we do 'concurrent everything.' We are constantly working in parallel; we collaborate with partners, customers, suppliers at every phase from product definition on."[25]

The diversification of local customer-supplier linkages further enhanced the capabilities of the cluster. Customer-supplier collaborations in a cross-Pacific IC supply chain that includes the manufacturers of semiconductor equipment, the foundries, IC designers, and PC and other electronic systems firms have contributed to Taiwan's growing share of the global IC industry. In 1999 TSMC and UMC jointly controlled 49 percent of the less than $3 billion global foundry market; by 2004, Taiwan's foundries jointly accounted for over 70 percent of the world pure-play market of over $18 billion (and the remaining competitors held only single-digit shares).[26]

Deepening and Diversifying the Supply Chain

Start-ups as well as established companies in Taiwan continue to diversify the local supply base by taking on the production of critical and new components. Some of these producers are very narrowly specialized. Waffer Technology, for example, an electronic component company, has developed a sophisticated process technology for making magnesium alloy casings for notebook computers. It boasts yield rates of 90 to 95 percent, well above the 70 to 80 percent standard in the industry. Another Taiwan-based specialist, Silitech, has developed deep expertise in the complex plastic and rubber composites used in manufacturing handset keypads; it has also brought together mechanical engineers, chemical process engineers, precision tooling, and diverse manufacturing processes in its ongoing efforts to design cost-effective keypads.[27] As in Silicon Valley, specialization allows innovation through focus for the individual firm, while also accelerating a dynamic process of innovation through recombination of these specialist components within the wider industrial ecosystem.

This mutual upgrading is clear in the partnership between Taiwan IC design firm Sunplus Technology and Silicon Valley's Silicon Image—a partnership that has allowed the former to transform itself from a source of low-end consumer electronics chips to designing high-end DVD and DSC chips. A 2003 agreement between the firms, for example, focuses on the co-development of multimedia ICs. Silicon Image brings leading-edge multimedia and storage-systems intellectual property to the partnership, while Sunplus excels in developing ICs for intensely competitive mass markets. The firms' engineers will jointly incorporate the Silicon Image IP into new products that both companies will sell; Sunplus will also develop lower-cost versions of Silicon Image's current chips to compete in commodity markets. The agreement is not exclusive: Silicon Image has also licensed its IP to two other Taiwanese IC design firms. In addition, Sunplus has an alliance with Oak Technology, another Silicon Valley–based firm, to jointly develop low-cost chips for China's imaging and digital TV markets and to establish a spin-off to develop optical storage ICs for the PC market.[28]

A firm that exemplifies the deepening capabilities of Taiwan's leading manufacturers is Asustek. The firm manufactures more than one million motherboards monthly (earning gross profit margins of 18–19 percent from this half of its business) and produces notebook computers for high-end brand name customers like Sony and Apple. Its strategy is to concentrate on the latest technology and to be first to market with motherboards that incorporate the latest design solutions. It can only do so, according to Joe Hsieh, director of the motherboard department, by working very closely with suppliers and customers. Hsieh reports that he regularly sends teams of engineers to work with PC manufacturers on quality and R&D—and that several of these suppliers have "grown up" with Asustek. The firm also cooperates closely as an alpha site with major chipset vendors like Intel and Taiwan-based VIA, often getting access to new chipset samples ahead of the competition in order to be early to market.[29]

Leading manufacturers like Asustek have the focus and supply-chain partnerships needed to adapt quickly to new market opportunities, as they did in the shift from low-margin desktop PCs to more profitable notebooks in the early 1990s. Taiwan's early move into wireless phones later in the decade provides another example of this responsiveness, as does its more recent shift into digital camera production. Figure 5.2 depicts the new combinations of specialists that evolved in Taiwan in the late 1990s to serve the emerging wireless phone market.

Taiwan's capacity for rapid imitation and adaptation reflects its excellence in technology implementation rather than a pioneering development of technologies or architectures. Producers like Asustek dominate the markets for technology products with three- to six-month design cycles and nine-month (maximum) life cycles in part because of their fast cycle times, tight inventory management, and adaptive supply chains, and in part because of their privileged access to leading-edge technologies, customers, and advanced research and

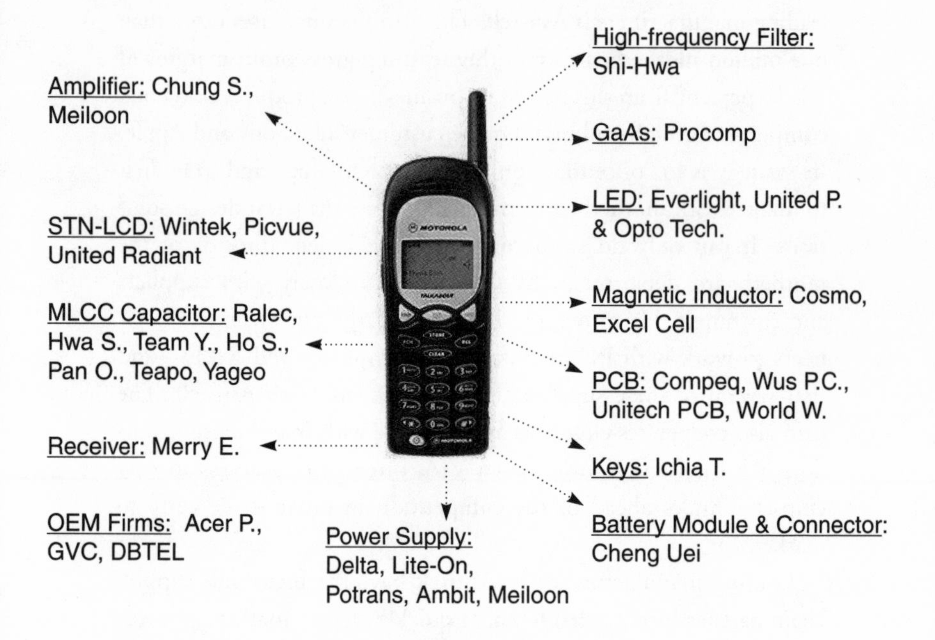

Figure 5.2. Taiwan's wireless phone industry. All companies are public.
Source: Wan-Pou Investment Company, Taiwan.

design from the United States. Over time, Taiwan's IC and PC firms have enhanced their own technological capabilities through their investments in research and development. In 2001 the companies in the Hsinchu Science Park collectively spent 7.3 percent of their sales revenue on R&D, compared to only 1.3 percent for Taiwanese manufacturing as a whole.[30]

The superior performance of Taiwan's manufacturers relative to U.S. contract manufacturers or electronics-manufacturing service (EMS) companies highlights the advantages of its decentralized ecosystem. A 2004 analysis estimated that Taiwan's top five ODMs—Asustek, BenQ, Compal, Hon Hai, and Quanta—turn their inventory 35 percent faster than the top five electronics-manufacturing service vendors (including Celestica, Flextronics, and Solectron) because the former have full responsibility for sourcing and supply chain management. This is reflected in an average return on equity for Taiwan's ODMs of 18 percent in 2002, compared to a negative average of 26 percent for the contract manufacturers. Likewise, analysts forecast a 34 percent average annual growth for the ODMs compared to 3 percent for the electronics-manufacturing service companies.[31]

While Taiwan's traditions of entrepreneurship, collaboration, relationship-based business, and resource sharing among small- and medium-sized producers have provided fertile ground for many aspects of Silicon Valley management models,[32] other customs, such as the heavy reliance on family ties, have largely been abandoned. As a result, Taiwanese businesses are often far more comfortable than their Asian counterparts setting up branches in Silicon Valley. This does not mean that Taiwan is a copy of Silicon Valley; rather, it has co-evolved with the California region in a hybrid that combines the Silicon Valley model of venture capital and entrepreneurship with the older traditions of resource sharing and collaboration of Taiwan's small- and medium-sized producers.

Building Regional Advantage

Taiwan's experience might appear to be a classic case of the benefits of comparative advantage, with Taiwan specializing in IC and PC manufacturing and Silicon Valley leading in more advanced IC design and electronic system definition. However, the economic gains achieved from specialization and trade are limited compared to the far more dynamic process of cross-regional collaboration and upgrading. The Taiwanese technical community—including engineers firmly rooted in the United States and those based in Taiwan, as well as those who constantly travel back and forth—efficiently transfers information about investment and partnership opportunities as well as about markets and technology between Silicon Valley and Hsinchu.

Whereas models of comparative advantage assume relatively stable industrial specialization of locations based on factor endowments, Silicon Valley and Hsinchu have continued to transform each other—and their industrial specializations and capabilities—over the past two decades. The boundaries between firms, industries, and even regional economies are blurring as the specialization of production, the co-design of products, and their rapid recombination generate new technologies, new products and processes, and new industries. These regions and their industrial specializations continue to evolve as producers engaged in joint innovation across geographically far-flung and protean supply chains.

TAIWAN AT THE TURN OF THE CENTURY

The Hsinchu industrial system, which successfully adapted to both competitive and technological changes in the 1980s and 1990s, faced its most severe challenge with the movement of manufacturing capabilities to the Chinese mainland as the decade drew to a close. Continued downward pressure on prices and shrinking gross margins resulting from cutthroat competition in the increasingly commodity-like PC business led Taiwanese firms to begin manufacturing in China. Their goal was to increase economies of scale (and capacity) while lowering production costs, and thus increase market share and profitability. However, with all firms pursuing the same strategy, they soon faced overcapacity, and even though computer notebook and other electronic system manufacturers gained market share, their margins and profits continued to fall.

China: New Pressure on the System

In the early 1990s Taiwan's PC component makers, driven by intensifying cost competition, began locating their most labor-intensive activities—the assembly of power supplies, keyboards, and scanners—in China to exploit the lower-cost labor and land. Following the path of an earlier generation of Taiwanese manufacturers of footwear, toys, and light consumer goods, they began to relocate to the south of China. What started as a trickle of labor-intensive manufacturing accelerated throughout the decade, and ultimately included even Taiwan's most sophisticated IT producers.

A majority of the early Taiwanese PC-related investments were clustered in the city of Dongguan, located between Shenzhen and Guangzhou in the Pearl River Delta. This provided proximity to Hong Kong's communication and transportation links, even though Guangdong is not the obvious location for Taiwanese investments. Fujian province is both geographically the closest to Taiwan (directly across the Straits) and also the closest culturally (most people who moved from the mainland to Taiwan were from Fujian). Taiwanese regulation did not permit travel directly across the Taiwan Straits, however, so all business traffic passed through either Hong Kong or Macao, and investments were typically routed through Hong Kong to avoid government limits on investment in the mainland. The rapid urban and industrial growth of Shenzhen, resulting from its status as the first Special Economic Zone in China, made it less attractive as a manufacturing center than peripheral areas such as Dongguan and Huizhou, where land and labor costs were significantly lower.

By the turn of the century there were some 1,800 PC firms clustered in the city of Dongguan, 80 percent of which had been established by investors from Taiwan, along with the production lines of major multinationals and a handful of domestic Chinese firms. This agglomeration included producers of 95 percent of the parts and components required to make a PC. However, since these were the most mature commodity products, the value of parts and components produced in China accounted for only 10 to 15 percent of total PC costs.[33] Standard keyboards, for example, sell for about $4.00 each, about half of which is the cost of materials.

By 2000 one-third of Taiwan's IT products were manufactured in factories located in China. This included 90 percent of all power supplies, 85 percent of scanners, and over 42 percent of all digital cameras, desktop computers, motherboards, and monitors.[34] Investors from Taiwan compare the density and sophistication of Dongguan's PC manufacturing infrastructure to that of the Hsinchu-Taipei corridor in the early 1990s: a decentralized and flexible system for low-cost mass manufacturing, with at least three-quarters of output destined for export.

Taiwanese IT investments in China accelerated at the turn of the century and began to shift toward higher-value-added products, including motherboards, video cards, and notebook PCs. Figure 5.3, which tracks the growth of Taiwan's IT output in the 1980s and 1990s, shows the fast-growing share of production overseas, overwhelmingly in China. These higher-value-added Taiwanese investments were also increasingly located in Shanghai and the neighboring cities of Suzhou, Kunshan, WuXi, and Hangzhou (Zhejiang and Jiangsu provinces) rather than in southern China.

This shift to Shanghai reflected the declining importance of proximity to Hong Kong for doing business, as well as aggressive promotion of IT by the provincial government. The Shanghai region also offers better access to the China market and a significantly larger base of technically skilled university graduates than Guangdong, which grew in importance as more complex manufacturing processes were needed for notebook computers and similar products. These new production centers also required the training and deployment of

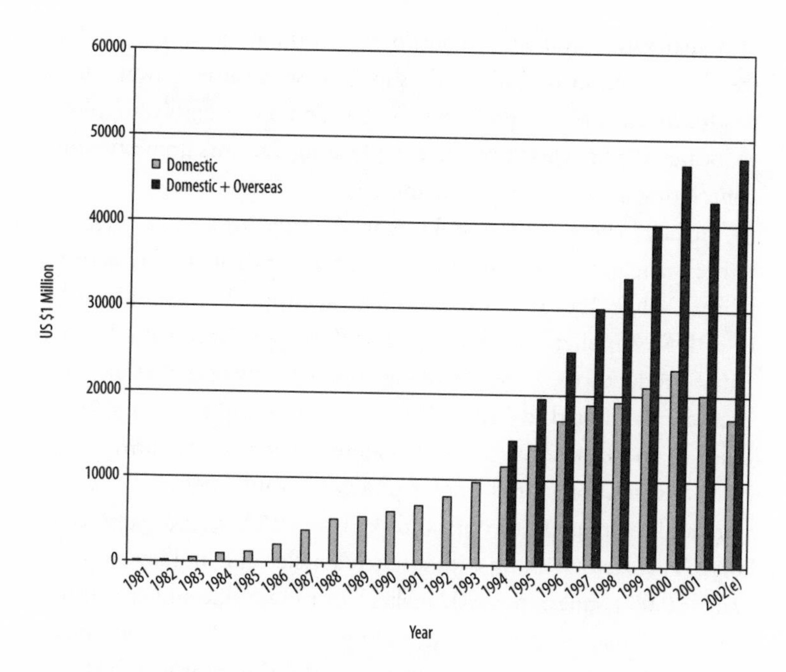

Figure 5.3. Annual value of IT production, Taiwan, 1981–2002.

substantially more Taiwanese managers, and even entire families, suggesting a deeper and longer-term commitment to doing business in the mainland. In 2000 Chinese officials reported that there were 3,000 Taiwan-based companies in the Shanghai area, along with more than 200,000 Taiwan-born residents.

The migration of IT manufacturing capacity across the Taiwan Straits was precipitous. By 2004, over 85 percent of the total production value of Taiwan's IT manufacturers ($63 billion in U.S. dollars) was generated in China, a 22 percent increase over 2003. (In 2004, 8 percent of production value was generated in Taiwan and 7 percent in other overseas locations, while in 2003, Taiwan accounted for 21 percent and other overseas locations for 16 percent.)[35] The transfer of higher-value production contributed to the deepening as well as the diversification of China's IT supply base because Taiwan's lead producers encouraged their component and subassembly suppliers to locate near final assembly, thus replicating Taiwan's domestic subcontracting networks on the mainland.

Virtually every Taiwanese IT manufacturer now has a presence in the mainland; many have moved the bulk of their employment to China. While Taiwan-based firms have continued serving as OEM or ODM suppliers for leading global corporations with brand-name products, they are managing supplier networks that are predominantly located in China.[36] Asustek, for example, in 2000 commenced operations in a 375,000-square-meter manufacturing facility in Suzhou for its full line of products, from motherboards and notebook computers to optical disk drives, PDAs, and game consoles. This factory, along with a much smaller one in Taoyuan, Taiwan (80,000 square meters), makes both low-value-added components and mature products in volume. Asustek has a very small facility (28,000 square meters) in Peitou, Taiwan, near its R&D and technical support departments, where it manufactures new and complex products that require active involvement of the engineering staff. The company also has small assembly facilities in the Czech Republic, Mexico, and Taiwan.[37]

Official data significantly undercount Taiwanese IT investments in China because the government limits the outflow of capital to China (the initial cap on investment of $50 m. per project was increased to $100 m. in 2002, with larger projects considered on a case-by-case basis). High-tech firms have historically avoided this restriction by registering in third countries (such as the Virgin Islands) or by routing investments through Hong Kong. However, even the official data reflect the dramatic growth of investment, from $875 million in 1997 to $1.5 billion in 2000 to $6.9 billion in 2002. Most analysts estimate the cumulative total of Taiwan's investments in China between 1989 and 2004 at about $80 billion.[38]

The loss of manufacturing and related service jobs as a result of this "hollowing out" of Taiwan's economy—combined with the economic downturn in the global electronics industry since 2001—led to a great deal of soul-searching about Taiwan's future among industry and policy leaders. The social cost of Taiwan's "fast follower" strategy, which combined rapid responsiveness, continuous engineering improvements, and cost reductions, became increasingly apparent when entire supplier chains could be relocated to China. These anxieties were amplified by the periodic eruption of political tensions with the mainland over Taiwanese independence. Historically, the strength of Taiwan's economy had been a major asset in maintaining a sense of autonomy and security in the face of a hostile power.

Government Adaptation

Taiwan responded to the Chinese challenge by conducting an interactive and decentralized dialogue to identify technological opportunities, research priorities, and strategies for collaboration. This policy dialogue, which became very widespread, grew out of the often ad hoc and incremental solutions to the collective challenges facing Taiwan's small- and medium-sized electronics producers and over time created a series of institutional mechanisms for consultation and coordination. The participants in this ongoing dialogue include industrial planning agencies, state-owned development banks, private venture capital firms, the leading Taiwanese electronics companies and affiliates of foreign companies, start-up businesses, university research centers, public research organizations, and the overseas Chinese business communities in the United States and elsewhere.[39]

Taiwan's science and technology agencies, and particularly ITRI's ERSO, helped to build a domestic IC industry in the 1970s and 1980s by transferring IC technology through a process referred to by local engineers as "R&D: reverse and duplicate." These agencies played a significantly less interventionist role during the 1990s.[40] In 2002 Taiwan was spending 2.3 percent of GDP on R&D, compared to 2.1 percent in Singapore, 2.2 in France, and 2.6 in the United States.[41] The result of Taiwan's multi-level policy formation process since 2000 is a set of coordinated strategies for upgrading the domestic economy, focused on three main objectives:

- Strengthening Taiwan's position as OEM/ODM supplier to global brand corporations in the face of competition from global contract manufacturers.
- Broadening capabilities in IC design, particularly system-on-a-chip (SOC) design, which combines computer memory, graphics, sound, networking, and CPU on a single chip.
- Creating value by developing global brands, intellectual property, and de facto industry standards.

Each of these objectives in turn involves a variety of different actors and institutional responses, including diversification into new product markets (such as digital video and audio systems), improvements in information management and in process technology, investments in specialized training and research oriented to SOC design, and establishment of a Global SOC Design and Service Park.[42]

Some observers believe that ITRI's greatest legacy will be as an advanced training center: between 1973 and 1998, approximately 10,000 ITRI researchers left the agency to join the private sector, taking with them the skills and knowledge as well as the technological capabilities that develop from working in an advanced research laboratory. The institute also claims that more than fifty of its alumni are now CEOs of publicly traded companies.[43]

In sum, the Taiwanese government's response to the challenge of the mainland has been technocratic, emphasizing education, research, and development. ITRI has adapted to Taiwan's evolving relationship with mainland China, focusing on stimulating creativity and technological innovation, upgrading capabilities in areas such as IC design and global logistics, and stimulating international research collaborations—including collaborations with mainland institutions.[44] For example, ERSO has invested in developing leading-edge technologies in the area of flat-panel displays, such as the carbon-nanotube field-emission display (CNT-FED), which promises to deliver better performance and lower cost than the currently dominant TFT LCD technology. This effort includes extensive collaboration with local materials companies, universities and research institutes, and established optoelectronic companies.

For Taiwan, these developments mark a significant shift away from its role as a technology follower. According to Dr. C. C. Li, deputy director of flat-panel display development at ERSO, "The IP we have developed at ITRI for the design and production of CNT-FEDs competes directly with that of South Korea, Japan, and the US. Normally at ITRI we are a little behind the leading institutes and companies, but in this case, CNT-FED development, we have developed simultaneously with other organizations and countries . . . If any company in Taiwan wants to develop and mass produce CNT-FEDs, we are confident that they would not have to worry about IP issues as they have had to do before, in the cases of PDP and TFT-LCD production, for example."[45]

Climbing the Value Chain

While it is too early to judge the success of the government's strategy for Taiwan, private industry—both foreign and domestic—is already adapting to the China challenge. Two trends are clear: Taiwan is becoming an international R&D center, and domestic producers are continuing to move into lucrative new industries and market niches—often by deepening their collaborations with key customers and suppliers.

Foreign companies set up research centers in Taiwan in the early 2000s at an unprecedented rate, even as they moved headlong into China. Seeking to be closer to key suppliers and to capture tax breaks and subsidies, more than a dozen leading technology companies established new international R&D centers in Taiwan by 2005, including Intel, IBM, Sony, Dell, 3Com, Broadcom, Conexant, HP, Motorola, Alcatel, and Nortel Networks. Corporate decision-makers emphasize the value of proximity to their Taiwanese partners as a source of innovation. Broadcom, for example, established its system-on-a-chip (SOC) R&D center in Taiwan because of the close proximity to its leading-edge IC design and manufacturing firms. The center will have 125 engineers doing IC development for products with high integration capabilities and low cost. Their goal is to share new IP technologies with local partners in order to jointly develop new products more rapidly.[46]

Similarly, HP established a product development operation in Taipei so that its engineers could work directly with designers from their Taiwanese suppliers. HP executives claim that co-location has enhanced teamwork dramatically and that they can now introduce as many as fourteen new laptop models a year, compared to ten a year when the two groups were on opposite sides of the Pacific. They are also getting each model to market much faster. The managing director of the HP operation says that "R&D in Taiwan can get you closer to the design phase. You don't see that kind of innovation in China yet."[47]

Nor are domestic firms abandoning Taiwan. As they shift volume manufacturing and product development to China, they are focusing in Taiwan on higher-value research, logistics, global business operations, and marketing. For example, Quanta Computer, the world's largest notebook manufacturer, has built a vast Quanta Shanghai Manufacturing City (QSMC) for its labor-intensive, high-volume production (taking advantage of wages that are one-sixth of those in Taiwan). The QSMC employs 20,000 workers in a 7-

million-square-foot manufacturing plant, with a 1.8-million-square-foot dormitory for workers, ample space for its key suppliers, a development lab, schools, and a hospital. At the same time, Quanta built an extensive new R&D complex outside of Taipei, designed to house 7,000 engineers who will design advanced notebook PCs, servers, and LCD TVs, and will oversee worldwide logistic services. Quanta's founder, Barry Lam, explains: "We have to do advanced research . . . more and more customers need us to design the whole thing. The market is so competitive. There are so many products that are similar. So we are forced to invest in innovative research in new products that are one or two years ahead of the market." Foxconn Electronics is also building a new R&D center in Taiwan to focus on development of nanotechnology used for molds and machinery, thermal technology, batteries, monitors, and optoelectronics.

For Quanta, Foxconn, and many other Taiwanese firms, the China activities are an inseparable extension of those in Taiwan—and the two are managed as an integrated entity. While manufacturers will continue to take advantage of lower costs in China, the superiority of Taiwan's managerial and marketing capabilities means that they will continue to maintain control over the production process in the foreseeable future.

The rapid expansion of optoelectronics production has fueled Taiwan's IT growth since 2000. By 2003 Taiwan's firms had gained 65 percent of the global LCD monitor market, thus reducing dependence on Japan for a key, and very costly, component. The Southern Taiwan Science Park—established in Tainan to serve as a counterpart to the overflowing Hsinchu Science Park in the north—has been dubbed "Crystal City" for its fast-growing collection of highly capital-intensive flat-panel manufacturers. Two of Taiwan's leading TFT-LCD firms—Chi Mei Optoelectronics and Hannstar Display—moved to the southern park, bringing dozens of upstream materials and parts makers with them. These optoelectronics firms accounted for 66 percent of the total park revenues of $8.82 billion in 2004. The Hsinchu Science Park also reported record revenues of $34.5 billion in 2004.[48]

Taiwan continues to spawn new market niches—which are often destined to create new industries. Many firms are exploiting and further developing local capabilities in a range of small-form-factor (SFF) devices. Shuttle Computer, which began manufacturing motherboards in 1983, pioneered the development of SFF PCs with its introduction in 2001 of the XPC, which combined a very compact motherboard and chassis with the capabilities of a desktop PC. The initial success of the XPC allowed Shuttle to exit the unprofitable commodity motherboard business altogether and focus on problems related to the SFF, including improving acoustics, aesthetic design, and accessibility. The firm has even designed its own LCD monitors because the monitors need to be rugged and high-performance while also remaining portable and stylish.[49]

Taiwanese suppliers also produce the key components for the new small-form-factor hard disk drives, or microdrives. These one-inch storage element (SE) devices, which are small, compact, low-cost, durable, and high-capacity (2–3 gigabytes), are enabling a new generation of "pocketable" consumer electronics products. These include USB personal storage devices, PDAs, digital audio players (MP3 players), digital video cameras, mobile phones, and GPS devices.[50]

Taiwan's producers are also exploiting the growth of wireless devices and the merging of the PC and consumer electronics sectors with products ranging from hand-held personal video players and wireless webcams to digital cameras and "media hubs"—wired or wireless boxes designed to channel audio, video, and still images around a networked home. The founder of BenQ, which stands for "bringing enjoyment and quality to life," seeks to create a branded consumer electronics firm that makes lifestyle products such as MP3 players, LCD-TVs, and digital cameras for markets in China as well as Europe and the United States. If BenQ founder and former Acer executive Kuen-Yao Lee creates a successful company out of the remnants of the former Acer Communications and Multimedia Division, he will be the first to succeed in this elusive goal of growing a brand from Taiwan.

Finally, Taiwan's IC foundry business is threatened by the emergence of new foundry competitors in China as well as the rising costs of maintaining technology leadership. The risks of excess capacity and price wars have intensified as the cost of a new fab has now surpassed $3 billion. The system-on-a-chip (SOC) products being adopted for communication devices such as mobile phones and broadband Internet equipment pose unprecedented design and technical challenges. Taiwan's firms are meeting this challenge by deepening their partnerships with chip designers and customers. TSMC, for example, now provides its leading customers with greater access to its engineering and manufacturing processes in order to avoid expensive production delays and technical failures. As discussed earlier, UMC deepened its partnership with Xilinx and invested in WIS Technologies, a Silicon Valley start-up that designs multimedia compression chips.

Some observers believe that Taiwan is poised to become a SOC "design foundry" service center in the coming decade—just as it assumed industry leadership in the 1990s as a provider of independent foundry services. TSMC's 2003 investment in Global UniChip Corp (GUC), a specialized design engineering service provider, confirmed the value of this new market niche. IC design and technology have become so closely intertwined with the rise of very complex SOC products (the chips are very large with large gate counts, creating ongoing signal integrity and power management challenges) that effective design requires intimate knowledge of the characteristics of the complex fabrication technology. Even testing and packaging are difficult for these extremely complicated designs. A specialized "design foundry" like GUC must be familiar with both design intellectual property (IP) as well as process technology in order to ensure that when a design goes into a manufacturing facility, it will produce sufficient yields in order to ramp up quickly.

Taiwan's IC industry is reinventing itself in the face of the competitive threat from China by upgrading its SOC design capabilities and foundry technology simultaneously. "Design foundry" services like those provided by GUC are emerging as another high-value-added segment of Taiwan's IC industry. The disintegration of its industrial system thus gave Taiwan's producers an advantage relative to less flexible Japanese and South Korean competitors by allowing them to move rapidly into the high-growth communications-related markets rather than the slower-growing market for standard PC chips.

The dynamism and adaptive capacity of Taiwan's decentralized IT ecosystem—which generated over $120 billion in revenues in 2005—remain unparalleled. At the same time that producers are shifting mass manufacturing to China, they are transforming Taiwan into technology-based service provider with sophisticated global integration and logistics capabilities. Taiwan's IC founders remain technological leaders and control 70 percent of the global foundry market.[51] Taiwanese producers also lead global markets for other IT and networking hardware products, including notebook PCs (73 percent), motherboards (78 percent), LCD monitors (68 percent), routers (89 percent), and WLAN products (83 percent).[52]

The new Argonauts transformed the global economy first by linking Silicon Valley and Hsinchu, and again as they tapped the skills and resources in mainland China. The U.S.-educated Chinese engineers who spawned Taiwan's IT industry in the 1980s began transferring technical and managerial know-how to the mainland a decade later. A new generation of overseas Chinese engineers in Silicon Valley in turn began returning home to promising entrepreneurial and professional opportunities, further accelerating the geographic relocation of IT production as well as the development of new institutions and new markets in China.

6 ❖ Manufacturing in Mainland China

Howard Yang was one of the first entrepreneurs to return to the Chinese mainland after working in Silicon Valley. Yang earned a Ph.D. at Oregon State University in the early 1990s and worked in the U.S. semiconductor industry for several years before returning to China to join the state-owned semiconductor manufacturing firm Shanghai Belling. Two years later Yang left to found China's first fabless semiconductor design firm, Newave Semiconductor Corporation, with two Chinese colleagues he had met in Silicon Valley (one was from Taiwan, the other from Hong Kong). Rejecting the dominant ownership model at the time, a joint venture with a state-owned enterprise, the team instead raised venture capital funding and started a private, Silicon Valley–style company.

Yang and his co-founders tapped their professional networks in the United States and Greater China to start Newave Semiconductor. The firm began operations in 1997 with financing from both Silicon Valley and Taiwan-based venture investors, as well as from China's state-owned Hua Hong Microelectronics. Newave was headquartered in Silicon Valley; its R&D, design, marketing, sales, and administration were in Shanghai; and the firm lined up Taiwan Semiconductor Manufacturing Corporation (TSMC) as its foundry. In a pattern that has become increasingly common, Newave was a global company from the start, leveraging the resources of three distinct and distant economies.

Howard Yang had few counterparts in the 1990s. Since 2000, however, the desire to contribute to the long-delayed development of China's economy has inspired growing numbers of engineers to return home. This return migration is reminiscent of Taiwan's experience a decade earlier. The economic downturn in the United States triggered interest in the Chinese mainland among a technical community previously reluctant to return home, and for the first time China appeared to offer promising professional and economic opportunities for scientists and engineers educated in the United States.[1] Several new developments—China's entry into the World Trade Organization, the announcement of the 2008 Beijing Olympics, and a surge of foreign investment—contributed to a "gold rush" mentality in the early years of the twenty-first century.

The brain drain has by no means been completely reversed. Chinese students still pursue graduate education in the United States in large numbers, but with much greater willingness to consider returning home. Among those who do return, a significant proportion do not remain. Migration is now a two-way street, and the circulation of skill, capital, and know-how between the United States and China has reached unprecedented levels. The overseas Chinese professional and technical community that built the bridge linking Silicon Valley and Taiwan in the 1980s and 1990s is now extending its networks to the technology regions in mainland China.

THE SILICON VALLEY–HSINCHU–SHANGHAI CONNECTION

The growth of a technical community that connects Silicon Valley with the coastal regions of China has created the third leg in a triangle of social and economic relationships between Taiwan, China, and Silicon Valley. As Chinese engineers extend their professional networks to the mainland, they contribute to the growth of an important new center of global IT production as well as deepening the division of labor between these increasingly specialized and interdependent regions. The Chinese state and multinational corporations are involved in the process, as are individual entrepreneurs, but it is the networks of overseas Chinese engineers with ties to Silicon Valley and Taiwan's Hsinchu region who have transferred the technical and organizational know-how as well as the business connections that are vital to economic success in the current era.

By 2000, more than two decades of labor-intensive, small-scale investment from both Hong Kong and Taiwan had created the infrastructure in coastal China for efficient, low-cost manufacturing of consumer electronics. Taiwanese investments in large-scale IT manufacturing complexes in the Guangzhou and Shanghai regions in the following years turned China into the world's third largest IT manufacturer. The transfers of technical and managerial know-how associated with billion-dollar IC foundry investments transformed Shanghai from a relative backwater to its current position—roughly one generation behind the global frontier in IC manufacturing.

At the same time, returning entrepreneurs from Silicon Valley are seeking to exploit their knowledge of the language and culture to gain a competitive advantage in the Chinese market. They are starting new technology businesses in fields ranging from IC design to wireless communications to advanced ceramics. These entrepreneurs bring their know-how and their connections to Silicon Valley and Taiwan to a competitive landscape that is still largely dominated by state-owned enterprises.

The overseas Chinese technical community has thus twice altered the geography and structure of the semiconductor industry—first when engineers returned from Silicon Valley to Hsinchu in the 1980s and 1990s, and more recently when they moved to China, from across the Taiwan Straits and from Silicon Valley. Taiwan's semiconductor industry originated with talent and technology transferred from the United States, but when TSMC pioneered the IC foundry, it transformed the structure of the industry and accelerated the pace of innovation in both regions. The United States became home to the world's most sophisticated IC design firms, and Hsinchu to its leading foundries.

Technology and talent have moved once again—this time to the Chinese mainland. China's engineers are now managing dozens of IC foundries and at least 500 IC design ventures—most with trailing-edge technology and very low cost structures—and the growing market in China provides opportunities for experimentation in fields ranging from logistics to wireless communications. Much of this innovation is targeted at the needs of the Chinese market, and thus is potentially applicable to other developing economies as well.

HIGH-TECH MIGRATIONS

China replaced Taiwan as the world's third largest information technology producer in 2002, following the United States and Japan. The more than $92 billion in IT exports from China included over a third of the world market for mobile phone handsets, personal computers, and other electronics, computer, and telecommunications products ranging from digital still cameras and DVD burners to PDAs. However foreign firms, most from Taiwan, accounted for some 85 percent of these exports.[2] Since 2000 Taiwan's leading IT companies have built vast advanced manufacturing complexes in China, which in term became magnets for domestic investments. Foxconn (Hon Hai Precision) and Quanta Computer, for example, each exported over $8 billion worth of IT products from China in 2004. As in the United States, this IT activity was highly localized.

Three regions—Guangdong, Shanghai, and Beijing—account for 65 percent of Chinese IT output and 84 percent of IT exports.[3]

Taiwan's labor-intensive electronics manufacturers had moved to the southern provinces of Guangdong and Fujian in the early 1990s to take advantage of low-cost labor as well as aggressive recruitment by local officials. By the turn of the century, manufacturers of more complex, higher-value electronic systems like notebook computers and peripherals were making large-scale investments in the Yangtze River basin near Shanghai and typically included key suppliers when they moved. As a result, entire ecosystems for IT production were created in the greater Shanghai area within a very short time span.

Shanghai also attracted two billion-dollar high-profile IC foundry investments in 2001 in a development that triggered the migration to Shanghai of the entire IC supply chain—including equipment manufacturers, designers, wafer and other input suppliers, and packaging and test specialists. These investments were managed by the same generation of engineers, and in some cases by the same individuals, who had built Taiwan's semiconductor industry a decade earlier. The construction and operation of the new IC foundries involved what is perhaps the largest transfer of managerial and technical skill in human history. This process dwarfed earlier Taiwanese investments in China and was designed to transfer the tacit knowledge required to attain acceptable yields in the very complex and context-sensitive IC manufacturing process.

At about the same time, U.S.-educated engineers from China began to return home from Silicon Valley in large numbers. Most preferred to work for multinationals or to start firms that allowed them to take advantage of the networks and capabilities they had developed abroad. In general, the returnees avoided working for domestic companies, particularly state-owned enterprises. This return migration, although distinct from the Taiwanese investments, reflected growing confidence among overseas Chinese in the pace of economic reform on the mainland, as well as the assumption that they were ideally positioned to serve the large China market.

Government planners and businesses in China were acutely aware of the role that the reverse brain drain had played in the growth of Taiwan's IT industry. They were determined to take advantage of the post-2001 technology recession in the United States, and they competed aggressively to recruit returnees. Official delegations from the eastern Chinese cities that were already technology centers (Guangzhou, Shanghai, and Beijing) visited Silicon Valley regularly, hosted by the local Chinese professional and technical associations. Indeed, most of these SV associations soon opened mainland Chinese branches. These post-2000 returnees from Silicon Valley preferred overwhelmingly to move to Shanghai. The contributions of these fast-growing cross-Pacific networks to the economy of China are potentially far-reaching. Still, the Chinese business environment remains daunting even for native-born Chinese familiar with local languages, institutions, and culture.

Technology Meets Nationalism

China has steadily introduced market mechanisms into its planned economy for more than two decades, but institutional change remains uneven and fragmentary. While market competition has increased and policymakers have invested heavily in both physical and technological infrastructure, China's financial institutions, regulatory system, and legal system remain largely unreformed. Privately owned firms are now officially recognized, although bureaucrats regularly privilege producers with strong government connections. Personal access to government decision makers is still an essential element of business success. In sum, while government control over the economy is being reduced, there are few indications that China is creating a Western-style laissez-faire market economy.

China's political leaders have regularly affirmed their commitment to national economic autonomy. Deng Xiaoping's often-repeated maxim that "Science and Technology are the Chief Productive Forces" reflects the importance of technology to the post-communist vision of a modern, powerful economy. The commitment to technology excellence provides the ideological justification for contemporary Chinese policy in regard to software and other high-tech industries. Naughton and Segal define this "techno-nationalism" as a policy orientation toward autonomy and independence from other states.[4] The original techno-nationalist states, like Japan and Korea in the 1980s, maintained strong central government control in alliance with large domestic corporations in order to create independent capabilities in critical technologies, as well as to develop institutions to diffuse those capabilities throughout the economy.

China may be pursuing a less aggressive variant of techno-nationalism, but—as President Jiang Zemin stated in 1995—the country remains committed to becoming "masters of our own fate." The market thus coexists with a state that provides preferential treatment to domestic producers. The sustained effort to establish domestic standards for technology products is the most visible evidence of this commitment to technological autonomy. The promotion of select producers or "national champions" with government contracts, preferential access to capital, and regulatory relief also creates opportunities for bureaucratic discretion as well as corruption.

In line with the commitment to technological autonomy, the Chinese government has invested aggressively in both research and technical education. China steadily increased its output of science and engineering (S&E) bachelor's degrees from 125,000 in 1985 to 330,000 in 2000, thus approaching the U.S. output in the same year of approximately 400,000; and the nation increased its annual output of S&E doctorates still faster, from 1,069 in 1990 to 8,153 in

2001—surpassing India's more stable output of 4,000 to 5,000 per year in the 1990s.[5] Research and development spending increased steadily during the 1990s as well to reach $10 billion in 2000, or 1.01 percent of GDP compared to only 0.60 percent in 1995. The Tenth Five-Year Plan set the goal of raising R&D spending to 1.5 percent of GDP by 2005, which is very high for a developing economy. By contrast, in 1999 India's R&D spending was 0.86 percent of GDP, Mexico's was 0.40 percent, and the ratio in most advanced economies is in the 2 to 3 percent range.[6]

Standards Wars

China's vision of technological self-sufficiency includes defining national technology standards independent of those developed or used elsewhere. The goal of the standards is to leverage China's large national market, allowing Chinese companies to develop intellectual property (IP) for the domestic market while positioning them to compete overseas as well. This policy is in large part a reaction to the fact that foreign companies control virtually all of the technology-related IP in China. The Director General of the Department of Electronics and IT in China's Ministry of Information Industry, Zhang Qi, argues: "Owning independent intellectual property rights and winning the initiatives in setting industrial standards should be top priorities for domestic manufacturers." To this end, the MII organizes industry alliances among domestic manufacturers to ensure that they are able to impose standards on foreign producers.[7]

Although there is no evidence to date that this strategy will succeed, it has created significant conflict between the government and foreign technology businesses. In 2000, for example, China's State Encryption Management Commission (SEMC) enacted a standard requiring all foreign businesses to register with the government any products containing encryption technology. This regulation was targeted at the growing dominance of the Windows operating system, along with other foreign technology products including mobile phone handsets and browser software. An international business and government coalition from the United States, Europe, and Asia succeeded, after a year of negotiation, in forcing China to rescind the plan and limit the registration requirement to products for which encryption was a core function.[8] Nonetheless, major Chinese ministries and agencies, along with hundreds of technical committees and industry associations, are developing domestic technology standards in a process that lacks transparency, planning, or channels for international input.[9]

More recent conflicts over encryption standards for wireless Internet security reflect these ongoing efforts. In 2004 the Chinese government announced a requirement that all wireless networking hardware in China must carry Wired Authentication and Privacy Infrastructure (WAPI) technology, a proprietary security standard that is incompatible with existing wireless infrastructure. Foreign companies were required to register with the government by a certain deadline and to work with one of a handful of Chinese companies that controlled the standard. The U.S. government objected vehemently to WAPI on behalf of a coalition of leading corporations; and only months before the deadline, Broadcom and Intel Semiconductor announced that they would stop shipping chips to China unless the government repealed the policy, which it eventually did in an unanticipated reversal that lifted the WAPI deadline indefinitely.

China remains committed to controlling access to its market and privileging domestic producers through the use of national standards. The MII has developed a Chinese mobile telephony standard called TD-SCMA for third-generation wireless networks, intended to replace both U.S. (TDMA) and European (GSM) standards. This standard is likely to become a significant part of the country's wireless infrastructure in the transition to the third-generation networks with their higher speed and capacity. China is also developing its own enhanced video disc (EVD) standard as a next-generation DVD system; and the Standards Administration of China has mobilized a working group devoted to creating Chinese standards for tags and other radio frequency identification (RFID) technology.[10]

Recruiting Overseas Talent

As we have seen, Chinese students pursued graduate studies abroad (mainly in the United States) in growing numbers in the 1980s and 1990s. By the late 1990s some 75,000 Chinese students were leaving each year to study abroad, accounting for an estimated 15 percent of all of China's postgraduate degree candidates.[11] While accurate data are scarce, the best estimates suggest that the return rate from the United States in this period fluctuated between 12 and 20 percent (significantly lower than the rate for most other countries). As one returnee who founded an Internet firm with two college classmates reported, "Over 80 percent of my classmates are working in the United States now, and very few have returned to China. Most of them are waiting for a good opportunity to return but it's very difficult because they don't know the China market."[12]

China's Fifth Five Year Plan (2001–2005) recognized the potential of the overseas community as a source of experienced managerial and technical skill and called for increased efforts to recruit qualified IT talent from abroad. Specific incentives are determined largely at the provincial level; a typical offer is Beijing's promise of permanent residency for the returnee's family as well as access to schooling for children, subsidies for purchase of the first home or car, stock equity awards, and low rents in incubators for returning overseas students.[13]

Most of the provincial governments in China, in order to reward experimentation and competition, provide "Returning Students Venture Parks" exclusively for enterprises run by returnees. By 2000 there were twenty-three such parks across China, and many other municipalities had policies to attract returning students. It is difficult to assess the impact of these incentives, but most evidence suggests that the expectation of greater professional and economic opportunity has been more important to returnees than these government incentives.[14] While the large domestic market and skill base are attractive to prospective returnees from Silicon Valley, the unpredictability of the institutional and regulatory environment remains a significant concern.

CROSSING THE STRAITS

In the 1980s and 1990s electronics and other light manufacturing industries from Hong Kong and Taiwan sought lower land and labor costs by locating in Pearl River Delta (PRD) cities, including Shenzhen, Guangzhou, Dongguan, and Huizhou. These were the earliest areas in China to open up to foreign investment, and they benefited from their proximity to Hong Kong. This first wave of investments involved electrical appliance and consumer electronics manufacturing (electric fans, hi-fi equipment, rice cookers, microwave ovens, and the like).

While policymakers in Beijing remained ambivalent about engaging with Taiwanese businesses, local-level officials recognized many benefits from actively nurturing Taiwanese investments. These included not only higher tax receipts and local growth, but also the ease of interaction that came from shared language and culture, as well as management practices, bookkeeping, inventory management, and gift-giving practices.[15] In some mid-sized Chinese cities, for example, Taiwanese merchants contributed between 30 and 50 percent of local tax revenue.[16] By 2000, the Pearl River Delta region was producing 79 percent of China's telephone output, 60 percent of its printers, 56 percent of fax machines, 51 percent of digital switches, and 23 percent of personal computers—with the most technology-intensive investments originating in Taiwan.

These investments created localized clusters of specialized inputs, components, and end-products. Dongguan, for example, is home to thousands of Taiwanese firms that specialize in manufacturing PCs and their components (power supplies, connectors, and so on) and peripherals (monitors, displays, keyboards, disk drives). Shenzhen, which has attracted multinational as well as Hong Kong and domestic investments, developed specializations in light industry and light assembly of telecommunications and computer-related products. IBM and Compaq both located their desktop PC manufacturing in Shenzen, and Dell has a manufacturing center for desktop PCs, notebooks, and servers in Xaimen in the Pearl River Delta. Acer has a PC assembly facility in Guangzhou, which is also home to manufacturers of both electrical and electronic products as well as software and business service providers. Huizhou is home to manufacturers of laser diodes, printed circuit boards, CD-ROMs, and batteries. These investments have drawn hundreds of thousands of unskilled workers from rural areas of China to work in manufacturing facilities, and the living standards in the PRD region are now among the highest in the nation.

China's manufacturing cost advantage remains impressive. A 2005 study by *Electronic Engineering Times* compared the cost of producing Dell's most popular color laser printer (the 5100cn, which boasts a very low cost per page and is priced below $1,000) in China and the United States. The study concluded that for an annual production volume of 250,000 units, the manufacturing cost in China was $578.90 per unit, or two-thirds of the estimated $953.50 per unit in the U.S. The most significant savings were in labor costs: $38.02 per unit in China versus $270.95 in the U.S.[17]

Clusters and Supply Chains

The geography of Taiwanese investment in China shifted significantly after 2000. Faced with unrelenting cost pressures from their international customers, Taiwan's electronic systems firms began to locate even high-value-added activities such as motherboard, video card, scanner, and laptop PC manufacturing in lower-cost China. Rather than remaining in southern China, they increasingly located their investments further north in the Yangtze River Delta, historically the wealthiest and most industrialized region of China. More than half of Taiwan's $6.64 billion officially approved investments in China between 1999 and 2003 were concentrated in Shanghai and the nearby Jiangsu and Zhejian provinces, which have quickly developed their own specializations as swarms of domestic companies emerged to produce low-end parts and components for Taiwan's OEMs and suppliers.[18]

Northern Zhejiang province is home to dozens of IC design and embedded software firms along with handset manufacturing (Hangzhou), as well as a range of IC packaging, testing, and low-end chip manufacturing operations (Shaoxing). Taiwanese investments in southern Jiangsu province have created clusters of computer peripherals producers as well as TFT, LCD, and packaging and testing (Suzhou), motherboard and notebook PC manufacturing (Kunshan), peripheral producers (Wujiang), and IC manufacturing, largely Japanese (WuXi).[19] Leading international PC firms such as Hewlett-Packard, Compaq, IBM, and Toshiba all have manufacturing investments in the Shanghai area as well as the Pearl River Delta. These districts are also the location of some of the most successful Silicon Valley returnee firms: UTStarcom, WebEx, and Alibaba, for example, all have major operations in Hangzhou.[20]

The geographic shift toward Shanghai was motivated mainly by the higher skill requirements of these post-1999 investments as well as a desire for enhanced access to the Chinese market provided by the cities' central location. The absence of flights from Taiwan (imposed by Taiwanese law) made it time-consuming and expensive for Taiwanese firms to do business in China because all flights from Taipei to Shanghai were routed through Hong Kong—meaning that it took the better part of a day to fly across 80 miles of ocean. Earlier, lower-value-added investments could typically be managed by one individual commuting from Taiwan, but the greater complexity of products like notebook PCs required the relocation of entire management teams to China.

The Shanghai area provided access to substantially greater supplies of technical and managerial skill than southern China possessed. The limited technical skill base in Shenzhen had forced firms to recruit engineers and managers from the north, many of whom were unwilling to stay for long periods of time in this newly urbanized area. Shanghai, by contrast, is home to several excellent engineering and management universities, including Fudan University and National Jiao Tong University;[21] it is second in the nation only to Beijing in the supply of university and advanced graduates. Taiwanese managers and their families, accustomed to life in Taipei, also preferred the cosmopolitan life in the greater Shanghai region.

A 2001 poll by the Taipei Computer Association found that 90 percent of Taiwan-based high-tech companies either had already invested or had plans to invest in the mainland. Many of these investments were not permitted under Taiwanese regulations, which strictly controlled investments in "strategic" sectors like advanced chip manufacturing. Although the ban on investments of more than $50 million was ended in late 2001 (replaced by case-by-case evaluation in a policy called "active opening, effective management"), channels for investing through holding companies in Hong Kong were well established. This means that the official estimate from

Taiwan's Mainland Affairs Council of $19.2 billion invested by 2001 (through 24,000 transactions) substantially understates the total cross-Straits investments.[22] Chinatrust Bank estimates that approximately 70,000 Taiwanese manufacturers had invested between $50 and $100 billion in China by 2002.[23]

The lower costs of land, labor, and facilities on the mainland allowed Taiwan's systems firms to build huge factories, substantially expand output, and further reduce costs through scale economies.[24] In 2004 the laptop maker Quanta, for example, produced 1 million units per month from its four plants in China, and expected to complete a fifth plant that would increase capacity to 1.3 million units a month.[25] In fact, the leading notebook computer makers (including Quanta, Compal, Inventec, Asustek, Wistron, Arima, and FIC) anticipated moving all notebook production to China by the end of 2005; their Taiwan offices would remain responsible for product development, technology research, material procurement, financial management, and marketing.

As Taiwan's laptop and electronic systems manufacturers shifted production to the mainland, most moved their entire supplier networks with them, recognizing the logistical advantages of proximity to their supply chain as well as their deep interdependencies. The brand-name customers for Taiwan's OEMs, such as IBM, Dell, HP, and Compaq (to whom they delivered more than $20 billion of products in 2001) demanded not only lower costs but also high quality and timely delivery of products.[26] Moving to China provided ample supplies of labor and land at low cost, but exposed them to substantial risks because local component and materials suppliers had dismal track records in terms of quality control and timeliness. This was in part because China had limited distribution and logistics services, and in part because of its hyper-competitive environment. Foreign companies regularly faced challenges when working with Chinese partners, including the loss of control over proprietary technologies and production methodologies (with limited judicial recourse) and the departure of key personnel to start "me-too" competitors with lower-cost/lower-quality knock-offs.

Taiwan's producers solved this dilemma by moving their entire supply chains en masse to China. Laptop manufacturer Quanta allocated ample space for suppliers in its immense Shanghai "manufacturing city." The chairman of Sunrex Technology, a supplier of keyboards for notebook PCs, explains why it is critical for component makers like Sunrex to support customers like Quanta: "Vendors, Taiwanese OEMs, and component suppliers rely heavily on each other. Although the vendors and OEMs want to lower component expenses, they need component suppliers to be profitable to maintain a vertically healthy production chain."[27] He adds that in 2002 Sunrex built a factory in Wujiang, Jiangsu province, to be near its main customers. Over time the firm increased sales and profitability by serving a growing base of local customers as well as its original Taiwanese partners, and by redesigning its products to optimize manufacturability. Logitech likewise helped its entire supplier chain, from ICs to cable wire and plastic mouse case manufacturers, to move to Suzhou because of the cost and quality advantages of an integrated local supply base from which to serve international customers.

The ecosystem of electronics manufacturing suppliers in China has in turn stimulated substantial new investments in logistics, supply chain management, and distribution-related software. Taiwanese corporations like Quanta are investing heavily in development of back office systems including Web-based applications for automating information reporting, decision support, transaction automation, accounting, design collaboration, distribution, transportation, and logistics.[28] These efforts have increased the overall efficiency of entire supply chains—lowering costs as they compress cycle times and reduce inventory. While this does not reflect breakthrough innovation, these incremental innovations in the manufacturing process are the source of the continually falling costs and improved quality and performance of IT products today.

Mainland Fever

The ease with which Taiwanese companies were able to locate reliable low-cost China-based manufacturing continued to be a tremendous advantage. By 2003 approximately 40,000 Taiwanese companies were located in the mainland, including some 10,000 in the greater Shanghai area of China, along with about 500,000 Taiwanese residents. The *Economic Daily News* (Taiwan) reported that the Taiwan Companies Association in Kunshan City, which had only 200 members in 1999, had tripled its membership to 600 by 2002. Best-selling books in Taiwan in these years had titles like *Immigrate to Shanghai* and *My Shanghai Experience.* While expatriate plant managers and engineers traveled regularly across the Straits to corporate headquarters in Taiwan, a range of social institutions—from separate kindergartens and high schools to Taiwanese-owned karaoke bars and restaurants—developed in enclaves like Kunshan City to support the families of this fast-growing immigrant population.

CHINESE CHIP MANUFACTURING

In spite of repeated governmental efforts to catch up, including joint ventures with foreign firms, China's IC companies remained a decade behind the world technological frontier throughout the 1990s.[29] This was true not only of state-owned IC manufacturing enterprises but also joint ventures with foreign companies, such as Hua Hong-NEC Electronics in Shanghai, Shougang-NEC Electronics in Beijing, and Motorola in Tianjin. Mainland Chinese engineers working in the United States regarded their home country as a technological backwater that would support labor-intensive manufacturing but not advanced IC design and production. When the founders of Silicon Valley's Chinese American Semiconductor Professionals Association visited the Institute of Microelectronics at the Chinese Academy of Sciences in 1991, they concluded that China was not yet ready for the IC industry and chose to focus instead on building business and professional connections in Taiwan. Another CASPA delegation that visited Shanghai in 1998 noted advances in the design skills and motivation of Chinese engineers but found few improvements in semiconductor manufacturing capabilities.

Likewise, a delegation from the North American Chinese Semiconductor Association (started by Silicon Valley's mainland Chinese rather than Taiwan-born engineers) was impressed by the level of R&D and the technical expertise of Chinese scientists and engineers but also concluded that "the overall semiconductor industry in China lags far behind"—largely because of its "inability to transform the technology to commercial success." Describing a research center in WuXi that had "the best facilities among all of the places that we visited, but was also the best example of what one should be concerned about in China," the members of the delegation noted the lack of managerial and production expertise and the limited understanding of markets and customers: "The outlook does not seem bright. In the semiconductor industry, the investment is so big, technology so sophisticated, markets so competitive, and development so fast, one has to be large enough in size to stay alive, let alone grow and profit. China has plenty of talented people and should have no problem [in attracting] competent engineers from overseas. The question is how to run a fab in China and make it profitable. This may be a question easy to answer, but a problem hard to solve."[30]

In 2001, two high-profile teams of investors solved the problem by mobilizing large-scale transfers of IC manufacturing and managerial know-how as well as capital and technology from Taiwan and Silicon Valley. The ventures—Semiconductor Manufacturing International Corporation (SMIC) and Grace Semiconductor Manufacturing Corporation (GSMC)—built IC foundries in Shanghai and quickly surpassed even established multinational investors with long histories in China, such as NEC and Motorola. SMIC attained a technological level in standard mass production that was only two years behind the world frontier, compared to the usual eight to ten years behind in the mid-1990s. By 2003, China's most advanced producers were producing 130- and 180-nanometer processes, thereby surpassing the 250-nanometer design standard that had been used by governments in Taiwan and the United States to distinguish sensitive or critical technologies that they were not willing to have transferred to China.[31] While China's IC industry is a technological follower and does not pose an immediate threat to producers in Taiwan or the United States, this accelerated upgrading highlights the speed with which managerial and technical expertise is being transferred by the overseas Chinese technical community to the mainland.

A Chinese government program designed to improve domestic capabilities in software and semiconductors, which are regarded as flagship or "pillar" industries, triggered this migration of skill and technology. State Council Circular 18, "Some Policies for Encouraging the Development of the Software Industry and the Integrated Circuit Industry," issued by the central government in June 2000, provided for generous tax reductions and rebates, access to government venture capital funds, foreign currency retention, infrastructure investments, and reductions in red tape for all IC enterprises in China, foreign as well as domestic.[32] The most controversial of the policies—a significantly reduced Value-Added Tax (VAT) for enterprises that designed or manufactured ICs in China—provided as much as a 14 percent cost advantage to IC firms located in China over those elsewhere. The time frame for depreciation of mechanical and electronic equipment was also reduced to three years from ten years for IC manufacturing and to five years for IC design.[33] Local governments competing to attract investment offered additional incentives, including discounts on real estate and subsidized loans.

Private Sector Solutions

SMIC and GSMC were the first companies to exploit the incentives in State Council Circular 18 by building manufacturing facilities in China. They transformed not only the Chinese semiconductor industry, but also competitive conditions in the industry globally. These privately owned and managed IC ventures represented a dramatic departure from the existing business models in the industry—state-owned enterprises or joint ventures with state-owned companies. Both SMIC and GSMC adopted the pure-play semiconductor foundry model and drew heavily on Taiwanese and Silicon Valley talent, technology, and capital. Both decided to base their operations in Shanghai, which had a demonstrated ability and willingness to attract foreign investment. Both companies built close ties with the Shanghai municipal government as well as the central government. While GSMC is entirely privately owned, its personal ties to China's leadership are visible. SMIC, now a public company, also reports minority ownership by several government-related entities.

GSMC was founded by two so-called "princes": Jiang Mianheng (son of China's former president Jiang Zemin), the firm's principal shareholder, and Winston Wong (son of the chairman of Formosa Plastics, Taiwan's most powerful business group), chairman of the board.[34] The firm is privately held and secretive, but it was able to raise $1.63 billion (in U.S. dollars) for its first foundry and, despite its aristocratic leadership, hired top semiconductor talent in operational leadership roles. President and CEO Nasa Tsai holds a Stanford Ph.D. in materials science, worked at Intel and National Semiconductor in Silicon Valley in the 1970s and 1980s, and was a founding member and senior executive at one of Taiwan's early semiconductor ventures, MOSEL. The executive vice president for technology development at GSMC, Dr. Been-Jon Woo, holds degrees from National Taiwan University and the University of Southern California (Ph.D.). She worked at Intel for twenty years, receiving twelve U.S. patents and numerous Intel technology awards, before moving to GSMC.

SMIC management and investors also have extensive experience in the Silicon Valley and Taiwanese IC industries. The CEO of SMIC, Richard Chang, holds an American Ph.D. and worked for twenty years at Texas Instruments before becoming CEO of Taiwan-based Worldwide Semiconductor Manufacturing Corporation (which was acquired by TSMC). SMIC initially raised about $1 billion in equity from Taiwanese and American private investors as well as semi-public investors from Shanghai and Singapore, and also secured $480 million in loans from Chinese government banks. The overseas Chinese investors who helped to develop Taiwan's IC industry are lead investors in these ventures. H&Q Asia Pacific's Ta-lin Hsu and Walden International Investment Group's Lip-Bu Tan—who established the first foreign VC firms in Taiwan—are both major investors in SMIC and hold seats on the board of directors. Other SMIC investors include Shanghai Industrial Holdings (17 percent), Integrated Silicon Solution Inc. (ISSI), Goldman Sachs, Vertex Management, Foxconn International (Hon Hai Precision Industry), SMIC management, Chartered Semiconductor, the Zhangjiang Park Development Corporation, and the state-owned software giant Beida Jadebird.[35]

SMIC management brings decades of experience from the Silicon Valley and Taiwan IC industries. Nine out of the company's ten most senior executives in 2002 held electrical engineering doctorates or master's degrees from U.S. universities, and each had between fourteen and thirty-two years of experience at companies such as Intel, Texas Instruments, and Micron Technologies.[36] SMIC also recruited approximately three hundred middle managers directly from Taiwanese IC companies and another fifty from leading American companies. The senior manager of fab operations, Jowei Dun, came from TSMC.[37] One of these recruits explains: "The salaries here [in China] are lower than they are in the U.S.—but there is a greater upside. Things are moving very fast here. SMIC built its fab in one year, which may be record time. There is tremendous room for growth in China."[38]

Learning from Taiwan

These two major investments appeared to vindicate China's shift away from the model of state ownership and limited foreign investment that prevailed in the 1990s. The new strategy, clearly influenced by Taiwan's experience, combined openness to private ownership with strong financial incentives in order to attract foreign investment—and with it technological and managerial capabilities. At the same time, China began privatizing the state-owned semiconductor research institutes and eliminated all tariffs on semiconductors. In 2002 an executive at the IC equipment manufacturer Applied Materials described the magnitude of these changes: "Just a year or two ago the typical manager in the Chinese semiconductor industry was a bureaucrat assigned by the government, and the entire sector employed perhaps 125 overseas managers and technical experts in total. Now there are 700 managers and engineers from Taiwan, the US, and Europe working at SMIC, 200 at Grace, 10 at ASMC, and whole teams come from UMC and TSMC."[39]

The Shanghai government played an aggressive role in this process. Despite moves toward private ownership, SMIC was financed initially with $480 million in debt from four major Chinese banks, and received $183 million in equity from Shanghai Industrial Holdings Ltd.—an investment and commercial arm of the Shanghai municipal government which remains SMIC's largest single owner. In 2002 the CEO and executive director of Shanghai Industrial Holdings was appointed to the SMIC board of directors. Two years later, when the business was publicly listed on NASDAQ, SMIC signed another $285 million loan agreement with the same four banks. At that time Lai Zing Cai, chairman of the board of Shanghai Industrial Holdings, took a seat on the board of directors.[40]

SMIC describes its management style as "completely Western" and emphasizes its strict IP protection. The business is registered in the Cayman Islands, which allows it to avoid Taiwan's limits on investment in the mainland, and it is also registered as a Delaware corporation, ensuring that the firm is covered by U.S. corporate and securities law and giving the company the option of raising capital in the public markets in the United States or Asia. SMIC management relies heavily on legal and financial advice from U.S. professionals, often overseas Chinese, and all SMIC employees receive stock options as part of their compensation. SMIC has also used the reputations and industry connections of its senior managers to establish partnerships with IC technology companies and research institutes in the United States, Europe, and Asia, expanding its technological capabilities and the range of devices it can manufacture.[41]

SMIC's rapid mastery of the complex process of IC manufacturing underscores how quickly the overseas Chinese technical community has transformed the global geography of the semiconductor industry. In 2003 the magazine *Semiconductor International* named SMIC's Shanghai facility a Top Fab of the Year for manufacturing excellence. The editor-in-chief wrote: "We based our decision on many factors, including manufacturing capability, ramp-time, degree of automation, contamination control procedures, and consideration given to worker and environmental health and safety. We found SMIC to excel in all of these areas. We were particularly impressed with how quickly SMIC was able to put copper processing capabilities into volume production."[42]

Technologically advanced producers like SMIC (with 12-inch, 300-mm foundries) remain exceptional in China today, with production overwhelmingly focused on the more mature, commoditized segments of the market. Analysts agree, however, that the global center of gravity in semiconductor manufacturing is shifting to Asia, and China in particular. The country is now home to a dozen dedicated foundries and over fifty IC fabrication facilities, and production is forecast to surpass $10 billion by 2007. One U.S. executive predicts that by 2015 half of all IC wafers in the world will be produced in China and Taiwan, compared to one-quarter today.[43]

Beyond Low-Cost Labor

Chinese policymakers attracted these large-scale investments by linking preferential taxes and other incentives (cheap land, subsidized loans, discounts on utilities, and logistical support) directly to production in China. Since labor accounts for less than 5 percent of the cost of manufacturing chips, the labor cost savings available in China are of limited value to foreign semiconductor firms, particularly relative to the risks of moving the extremely sensitive and complex fabrication process to the mainland. Moreover, during the 1990s domestic firms were only capable of supplying about 10 percent of the Chinese market. As a result, Taiwanese and other foreign manufacturers preferred to supply the mainland market from their more predictable home bases.

The incentives for IC manufacturing in China introduced by the State Council in 2000 transformed the competitive calculus in this capital-intensive and cost-sensitive industry. The additional 14 percent VAT savings, in particular, appears to have triggered the investments by both SMIC and GSMC, and enabled local producers to build market share by undercutting imports. While these cost savings would diminish quickly, the Chinese market—destined to be the world's second largest by 2010—remained a powerful incentive for locating in China. The differential tax policy was challenged by the U.S. government and the Semiconductor Industry Association as a violation of the World Trade Organization requirements, and in 2004 the Chinese government agreed to repeal the rebate for domestic manufacturers. By then, however, it had altered the competitive landscape.

The foundries had established a first-mover advantage in the Chinese market, particularly in serving the country's fast-growing electronic systems supply chains. Morris Chang, CEO of TSMC, who had publicly questioned the wisdom of investing in China for many years, in 2001 announced plans to build a foundry near Shanghai: "Our U.S. customers who have either joint ventures or wholly owned subsidiaries in China have indicated that sooner or later we would have to be there, because it costs them a lot in taxes to import our goods." The prohibition imposed by the Taiwan government on leading-edge investment in China has further crippled established foundries, TSMC and UMC, because SMIC exploits the lower costs of doing business in China as well as its privileged access to western chip designers desiring to avoid China's value-added tax.[44]

The intense competition among China's local governments to attract IC investments has led many observers to fear an industry-wide glut. One official reports that soon after SMIC and GSMC announced investments in Shanghai, the Beijing municipal government established a new incentive program for the IC industry that he referred to as "B=S+1" (Beijing = Shanghai plus one)—reflecting the intensely competitive incentive-bidding process among Chinese localities. A year later SMIC started construction of a 300-millimeter DRAM fab in Beijing, in a radical shift from its original plan, which had no mention of Beijing or DRAMs. Moreover, Beijing's water shortages and contamination as well as frequent dust storms make the city a very unappealing location for chip manufacturing, which requires purified air and water. In the words of one longtime IC industry observer from Taiwan, "That venture may seem to make little commercial sense, but the Beijing government really wants to keep ahead of the Joneses in Shanghai."[45]

The risk of overcapacity, already present in other capital-intensive Chinese industries such as steel and autos as well as household appliances and white goods, is compounded by Chinese policies that give the semiconductor industry strategic status. It has been further fueled by inflated foreign expectations about the China market. There is already excess capacity in low-entry-barrier businesses like mobile phones, where thirty new Chinese companies joined the fray in 2001 and 2002 alone. There is significant potential for oversupply and increased volatility in the semiconductor industry as a result of growing Chinese production, with the most intense competitive pressures already evident in the pure-play foundry business. SMIC, for example, has struggled financially for its entire history, despite having displaced Singapore's Chartered Semiconductor as the third largest foundry in the world in 2004. It still lags far behind Taiwan's foundries and cannot possibly invest in research and development at the level of these established competitors. TSMC, for example, has a $400 million research budget and over a decade of close collaboration with Silicon Valley's leading chip designers.

Building an IC Cluster

The unprecedented speed and scale of these international transfers of skill and technology transformed not only the Chinese but also the global IC industry. In less than five years the overseas Chinese technical community changed Shanghai into a global center of specialized IC design, manufacturing, and packaging, assembly, and testing firms with close links to domestic businesses, Taiwan, and the United States. Industry observers claim that the clustering of the IC industry in Shanghai together with the market, technology, talent pool, government support, and capital supply in China today resembles the state of development in Taiwan's Hsinchu region ten or fifteen years ago. Some even predict the industry will grow faster than it did in Taiwan because China has a large base of experience from the United States and Taiwan to tap and an existing model to follow.

The 2002 announcement by TSMC of its plan to build a foundry in the city of Songjiang, just outside of Shanghai, was part of a large-scale migration of IC foundries as well as both upstream and downstream producers to the mainland. Prominent chip design companies, equipment manufacturers, and packaging, assembly, and testing companies from Taiwan (including VIA, Sunplus, Global Unichip, Silicon Precisionware, and ASE) and Silicon Valley (Applied Materials, Avant!, Synopsis, ISSI, and SST) located facilities in the Shanghai area. By 2004, the Zhangjiang High-Tech Park was home to ninety-six IC companies, including three foundries, seven IC assembly and testing firms, seventeen equipment and materials vendors, nine R&D units, and twenty-two IC design (fabless) firms. The region increasingly offers the benefits of an integrated IC supply chain infrastructure not available outside of Taiwan and the United States.

China's presence in the semiconductor packaging, assembly, and testing (SPA&T) segment of the industry is as substantial as in its foundry operations, but less frequently noted, perhaps because of lower technology and skill requirements. The 78 SPA&T facilities located in China in 2004 accounted for nearly 18 percent of the world's facilities and 14 percent of worldwide employees—and constituted a larger share of the world's total than any other part of the semiconductor value chain. Like the foundries, the migration of

SPA&T facilities accelerated after 2000, motivated by the growing need for proximity to their customers, the electronics manufacturing services and OEM/ODM companies. This clustering of specialist firms reflects the fact that two-thirds of China's semiconductor output is designed into products and exported.[46]

Most analysts believe that the current constraint on growth of the Chinese semiconductor industry is experience in IC fabrication, management, and marketing; they estimate that China will need 50,000 semiconductor manufacturing professionals and another 250,000 IC design engineers to support planned industry growth.[47] While China surpassed the United States in 1997 as the world's largest source of university-trained engineers, Chinese universities graduate only about 1,500 microelectronics students annually, and their quality is significantly inferior to those educated in the United States. This quality gap is attributed to an education system that does not emphasize creative thinking and problem-solving skills, together with a lack of professors with experience in a commercial setting.[48]

It is therefore no surprise that the Chinese government continues to aggressively recruit overseas talent. The Shanghai Municipal Government, for example, sent an announcement to the members of the Silicon Valley Chinese Engineers Association, offering to fund research projects in any of a wide range of areas identified as critical to the development of Shanghai as an "international metropolis": physics, chemistry, new materials, genetic technology, semiconductors, electronic and electrical engineering, human resources, intellectual property rights, sustainable development, regional and urban planning, environmental protection, and many other areas.[49]

While China's foundries are predicted to account for 20 percent of worldwide capacity in 2006, they lag world standards in critical dimensions of performance including turnaround (or cycle) times and quality of service and manufacturing technology. Developing these capabilities will take time and experience, particularly given the technological complexity of next-generation IC fabrication processes and the integration challenges associated with IC design. Although the global IC industry has been permanently transformed by China's new capacity, the competitive advantage of the current industry leaders in the United States and Taiwan in IC design, semiconductor intellectual property, equipment manufacturing, electronic design automation, and leading-edge process technology remains secure.

SILICON VALLEY MEETS CHINA

At the same time that cross-Straits investments shifted toward electronic systems and ICs in the late 1990s, growing numbers of Silicon Valley–based entrepreneurs, investors, and professional service providers began building connections to mainland China. As the president of the Silicon Valley Chinese Software Professionals Association noted in the late 1990s, "Many of our members from the mainland want to go back to China now to duplicate the Taiwan experience. . . . They have the experience and the technology to really help China lift itself up by its own bootstraps."[50]

The traditional route home for overseas Chinese scientists and engineers was to work for a multinational corporation with a research lab or branch plant in China. The western salaries and benefits, along with training and access to the resources of a global corporation, make jobs in foreign technology firms like Intel, Microsoft, or Oracle in China appealing. While many Chinese graduates continue to pursue this route, in the 1990s a growing number of engineers, particularly from Silicon Valley, preferred to try their hand at starting their own businesses.

Hong Liang Lu, a U.C. Berkeley graduate and co-founder of

UTStarcom, one of China's most successful entrepreneurial companies, spoke for many when he observed that the opportunities for mainland Chinese in America were limited, and added: "In my roots I am always wanting China to be a better and stronger country. I live here [in the United States]. I pay taxes. I try to be a good citizen. But this country doesn't need my help as much."[51] Like their Taiwanese counterparts, returnees reported being influenced by the more attractive professional opportunities and social status they could attain in China.[52]

The experience of China, however, differs in important ways from that of Taiwan. There is a substantially larger community of mainland Chinese in the United States—many of them motivated to return by the opportunity to serve the China market. They have the benefit of more than a decade of Taiwanese investment in developing an infrastructure of electronics suppliers in the mainland, which is now available to support technology start-ups. Policymakers in China have learned from Taiwan's experience and have created attractive incentives to lure returnees, and there is a deeper experience base among the investors and entrepreneurs who helped to build Taiwan's IT industry.

On the other hand, China's incomplete transition from a planned economy has created a daunting environment for return entrepreneurs. The uneven progress of political reform in China is evident in a physical infrastructure that has dramatically outpaced the development of the institutional infrastructure of a market economy. China boasts fifty-three nationally designated new- and high-technology zones and hundreds of science parks with advanced infrastructure and aggressive promotional policies—but they lack the social structures, the technical and managerial experience, and the financial and legal institutions that support entrepreneurial success in Taiwan and the United States.

Beijing's Internet Boom

The technology boom of the late 1990s triggered a small but notable increase in the number of returnees from the United States to China, mainly to Beijing. This included individuals from very diverse backgrounds (and as likely to have experience in financial services or marketing as technology), all eager to be part of China's Internet boom. However, the great majority of Internet start-ups failed in China, as elsewhere. Andy Lee, a graduate of Stanford University and one of Silicon Valley's first successful mainland Chinese entrepreneurs, reported in 1999 that he had persuaded many friends to return to China to start businesses, but most of them had failed and were not even able to recoup their original investments. Lee's remarks reflect the general frustration at the time that the environment in China was not yet ready for returning entrepreneurs: "People in China look upon returning students primarily as investors, assuming that they have a huge amount of money in their pocket; while they do not attach much importance to their technology and value . . . There is no way for private firms to compete with their rivals on a fair ground in China."[53]

A year later Lee returned to China in order to spin off a business from the Software Institute of the Chinese Academy of Science in Beijing.[54] Lee was already independently wealthy and had a U.S. green card, however, so he was not taking a significant personal risk. For most early returnees to China, having the option to go back was a precondition for leaving the United States.

Most of the early returnee start-ups in China focused on markets that were either premature (e-commerce) or over-hyped (3G, Internet marketing). Few had business experience or knowledge of the China market. The highest-profile returnee enterprises of the 1990s were "me too" replicas of U.S. companies and business models. The Internet portals Sina.com, Sohu.com, and Netease.com, for example, received extensive publicity, quickly gained market share in China, and even went public on NASDAQ. The same returnees also learned that U.S. business models were rarely appropriate in the Chinese market, where advertising revenues were scarce and online payment systems, distribution, and bandwidth often nonexistent. They had difficulty generating cash, gaining access to distribution channels, and establishing barriers to competitors.

The returnees learned finally that China is not a single homogeneous market. A Hong Kong banker observed: "Everyone sees 1.3 billion people buying things online, but it turns out that 800 million of them are sitting on the farm."[55] China's rural poor subsist on less than $1.00 per day, while the remaining 500 million live, on average, on about four times that amount. Only 20 percent of the latter group earn as much as $2,000 per year, which means that significantly fewer than 100 million Chinese have truly disposable income to buy items beyond day-to-day basic needs. In addition, the country has limited distribution, differing (and sometimes conflicting) provincial rules and regulations, and different languages—all of which suggest multiple markets. China Mobile, for example, has different services, prices, and business plans for each of the provinces.

The Internet portals Sina.com, Sohu.com, and Netease.com did not make money in China until 2002, when they provided the short messaging services and other content that now account for most of their revenues. Even today their future remains uncertain. Despite billion-dollar valuations, they provide near-identical offerings for both advertisers and users, and they remain subject to the editorial whims of regulators. A 2004 ruling by China's State Administration of Radio, Film, and Television (SARFT) that prohibited advertisements promoting "fortune-telling" services, for example, radically reduced revenues for Sina.com, which had significant income from that category of ads.[56] Many observers believe Chinese Internet companies are significantly overrated.

The early returnees faced other, more personal challenges when trying to start a Western-style business in China. The resentments of their counterparts who had never left China (and did not benefit from preferential government policies) were reflected in the pejorative labeling of returnees as "turtles from the sea" (in contrast to the local "turtles from the puddle"). The competitive tension between these two groups made it difficult for them to cooperate and learn from each other, and it exacerbated differences in management styles. This was clear in the high-profile conflict between Zhidong Wang, the successful Beijing entrepreneur who followed the traditional Chinese family-run business model when he became CEO of Sina.com, and the company's U.S.-educated senior executives and shareholders from Silicon Valley. Many returnee entrepreneurs also faced costly conflicts between the locally hired managers and those recruited from Silicon Valley, since the latter were significantly better paid than their local counterparts in comparable positions.

These returning entrepreneurs faced frustrating obstacles from the government as well. China's Ministry of Information Industry (MII), for example, delayed approval of Sina.com's request to file a public offering on NASDAQ for so long that the firm missed the window before the stock market corrected, and it was forced to delay its public offering for several years. In 2000, the MII halted for review all deployments of UTStarcom's Personal Access Services (PAS) systems and handsets—essentially a mini-cellular network that works within a limited geographic area. The firm was later allowed to deploy the PAS, but only in small cities, towns, and villages because it was not technologically advanced. This MII ruling left the company, which continued to win and fulfill contracts in large Chinese cities without a license, in prolonged legal limbo.[57]

B2C (Back-to-China)

The second, and significantly larger, wave of returnees from the United States to China began after 2001. China's Xinhua News Agency reported in mid-2002 that some 130,000 Chinese who had studied overseas had come back in a "growing tide of returning talent."[58] Although these numbers cannot be verified, the general trend is indisputable. These returnees were typically experienced technologists and managers from Silicon Valley, and their destination was far more likely to be Shanghai than Beijing or the southern provinces.[59] Lured to China by the promise of economic and professional opportunities not available in the United States, this generation of returnees has pioneered technology start-ups in fields such as telecommunications, component manufacturing, and IC design—and it appears they will have a lasting impact on China's economic trajectory.

The combined impact of the technology recession and the events of September 11, 2001, in the United States provided strong inducements for mainland Chinese working in Silicon Valley to consider returning home. A major policy reversal that called on overseas Chinese to "serve the nation" also signaled changing attitudes toward returnees, who were no longer viewed as "class enemies" if they failed to return.[60] A Chinese career search Web site reported a dramatic increase in résumés in early 2002: "In the past couple of months we have about 10,000 to 15,000 people interested in returning to China to work. Six months ago we would get only 10 to 100 résumés in the same period."[61] This trend, which was quickly dubbed "B2C" in Silicon Valley, was described by a former Microsoft employee who returned to start a company: "It's a big thing, now people just want to return to China. It's like the Gold Rush. They're successful in the US, but in their hearts they still feel like immigrants. They feel welcome here in China . . . there are not many new opportunities in the US."[62]

The speed of this change was striking. In the 1990s, Chinese technology companies were unable to recruit Silicon Valley engineers to return to work in China's IC industry, even when they offered to pay U.S. salaries. In 2001, however, a delegation from the Shanghai Pudong District attracted more than a thousand potential recruits to a session in Silicon Valley. The delegation, which included executives from thirty-five Shanghai businesses along with political officials, reported receiving more than 2,500 U.S.-based applicants for the 238 positions available. While these numbers should be taken with a large grain of salt because they were supplied by China's state-run Xinhua News Agency, the individuals who attended the event were taken aback by the unprecedented turnout and level of interest in jobs in China.

The two-way traffic of engineers and investors between the United States and China increased dramatically during this period, with Silicon Valley's Chinese professional and technical associations leading this accelerating process of "brain circulation." To cite a few of many examples: the U.S.-based Chinese-language newspaper *Sing Tao Daily* reported in 2001 that four of the first nine chairmen of the Chinese American Semiconductor Professionals Association (CASPA) had moved to Shanghai, comparing this change to the "tide of returnees to Taiwan" in the 1990s. Virtually all of the overseas Chinese professional and technical associations now have active branches in one or more cities in China, whereas none existed a decade ago. CASPA now has branches in Shanghai and Guangzhou in addition to Taipei and Silicon Valley. Even the Asian American Multitechnology Association (AAMA), founded in 1979 with a domestic U.S. focus, has announced the formation of a branch in Shanghai.

CHINA'S SILICON VALLEY?

The Central Committee of the Chinese Communist Party sought to encourage entrepreneurship and market competition in the 1990s as a means of accelerating the development of domestic science and technology industries. Influenced by the experience of the West, and Silicon Valley in particular, they invested heavily in new high-tech industrial zones to provide an environment conducive to innovation and entrepreneurship. The confirmation of the legitimacy and status of private firms as agents of economic development at the 15th Party Congress in 1997 advanced this process by eliminating the preferential treatment granted in the past to state-owned enterprises. All technologically advanced domestic enterprises (designated by the government) are now officially granted the same preferential treatment, regardless of ownership.

China's Ministry of Science and Technology (MOST) has designated the location of "new and high-technology industrial development zones" where the nation's resources are to be concentrated to support technological development. However, the purpose of the zones remains confused by multiple goals, differing strategies and tactics, and lack of an explicit vision of the desired outcomes. It is not even clear whether they are designed to attract technology to China or to encourage the technological development of Chinese firms.[63] While created by the central government, these zones are one of the prime targets of the competition between provinces and municipalities, which control everything from land ownership and access to the local market to the financing and regulation of start-ups.

Not surprisingly, the two leading zones in China differ along many dimensions (as they do from all of the others). The older Zhongguancun Park in Beijing has spawned far more indigenous technology businesses, while the Zhangjiang Park in Shanghai has attracted the lion's share of the technology-related foreign investment. Zhongguancun, often referred to as "China's Silicon Valley," was established in 1988 in an area that boasts China's two premier universities, Peking University and Tsinghua University, as well as 54 other universities and 232 research institutions, including the prestigious Chinese Academy of Sciences. The park claims to be home to more than 6,000 enterprises, of which 80 percent are in the IT industry. Zhangjiang Park was established by MOST in 1992 as the second national center for the development of new and high-technology industry. With 10 square kilometers of developed area, it is several times the size of Taiwan's Hsinchu Science Park, and it reports a population of 383 establishments.[64]

Planning Entrepreneurship

China's initial experiment with technology start-ups in the early 1980s encouraged would-be entrepreneurs to spin their enterprises off from local universities and research institutes. These were dubbed nongovernmental *(minying)* or "collectively owned" enterprises because their initial technology, offices, and equipment came from the state, but the firms were established and operated as independent entities. The Legend Group (a spin-off from the Chinese Academy of Sciences that was recently renamed Lenovo) and the Founder Group (a spin-off from Peking University) are two of the leading successes of this generation of nongovernmental technology firms.

This first generation of Zhongguancun start-ups—Legend, Founder, Great Wall, and Stone—are now among China's largest technology companies, with substantial output and employment. Legend, which was founded in 1984, had some $3 billion in revenue and more than 10,000 employees even prior to its 2005 acquisition of IBM's PC unit. These firms dominate China's PC market, with Lenovo at 29 percent, Founder at 9.2 percent, and Great Wall at 4.6 percent; in comparison, the most significant outside player was IBM at 5.3 percent.[65] The fact that these giants have not spawned a subsequent generation of successful start-ups after more than two decades underscores the limits of the Chinese environment for entrepreneurship.

The first-generation technology firms benefited extensively from their connections to the government, despite their designation as "nongovernmental." Students of China's technology sector refer to these famous university spin-offs (including Lenovo, Beijing Beida Jadebird, Tsinghua Tongfang, Shenyang Neusoft Group, and the Sichuan TOP Group) as the "new state-owned enterprises" because all of them have benefited from substantial and ongoing governmental support. These firms have also aggressively pursued growth, often through diversification into unrelated fields like real estate, following the government's desire to promote a small number of global "flagship" giants in each industry. Many have become conglomerates, failing to recognize the business advantages of specialization and focus. Beida Jadebird, a software spin-off from Peking University, for example, acquired a Hong Kong department store when the company was preparing to seek public listing in China. Lenovo likewise has extensive activities in real estate, investment banking, catering, and insurance industries, in addition to PC manufacturing and distribution, through its venture capital arm. As long as there is no market penalty for such unrelated diversification, this behavior is likely to continue. Even high-profile Chinese multinationals like Huawei and Haier are far more diversified than their U.S. competitors.

In spite of the thousands of start-ups they house, China's high-technology parks have largely failed to incubate successful private technology businesses. Although the Zhongguancun High-Tech Park boasted thousands of firms in 2004, the great majority were quite small and financially weak, apparently subsisting only on government subsidies for park ventures. A handful of very large companies account for the majority of total park output.[66] One Chinese engineer concludes: "Most high-tech promotion in China is just old wine in new bottles—an attempt to grow entrepreneurship using the tools of the planned economy."

This failure reflects one of China's central economic predicaments. The private sector in China accounts for 43 percent of all registered enterprises but only 13 percent of retail sales, 12 percent of industrial output, 8 percent of employees, and 4 percent of exports. In 2002, the average number of employees per enterprise in the private sector was 14, compared to 188 in state-owned or controlled enterprises and 36 in foreign-invested enterprises.[67]

Financing the Future

Private firms in China are severely disadvantaged relative to their state-owned counterparts because of their limited access to capital. The Chinese financial system is still dominated by state-owned banks, which are oriented almost exclusively toward state-owned enterprises, even those with minimal or negative returns. Central Bank data show that in 1999 less than 1 percent of all working capital loans went to private companies, and these were generally not to local private firms but rather to joint ventures with foreign private companies.[68] This problem is particularly acute for technology firms because commercial banks in China have limited credit-analysis capabilities and prefer to lend against tangible assets. Very few companies have publicly available credit records; most debt cannot be securitized, and therefore the great majority of China's new technology enterprises are underfinanced.[69] As a result, local start-ups cannot compete with foreign and large domestic firms for the best technical and managerial talent.

A key problem for the Chinese financial system that affects the entire economy, including the technology sector, remains the overwhelming burden of nonperforming loans (NPL), which in 2002 accounted for 50 percent of total banking sector assets. The total cost of cleaning up the NPL problem, according to Standard & Poor's, would amount to $700 billion, or about 50 percent of China's current GDP. Given the country's lack of financial management systems, skills, or infrastructure, full resolution of NPLs and reform of China's financial system would appear to be at least a decade away. The development of a healthy financial system depends, of course, on a reliable and well-functioning legal system to enforce financial claims and property rights. Reform of China's sometimes corrupt and often unreliable legal system goes to the heart of the democratic rights of its citizens, but it does not appear to be at the top of the government's domestic policy agenda.

Government "Venture" Capital

The Chinese government initiated the creation of a domestic venture capital industry in the 1980s as a mechanism for supporting new technology ventures. But in contrast with Taiwan, where a dynamic private venture capital industry took off in the late 1980s after a strong initial push from the state, China's venture capital industry emerged in the context of the state-dominated financial system and remains constrained by a largely unreformed capital market inherited from the planned economy.

China's first experiment with venture capital was the China New Technology Venture Investment Corporation, established as a limited corporation in 1985 by the State Science and Technology Council and the Ministry of Finance. It was declared bankrupt and closed by the People's Bank of China in 1997. This early failure exposed structural problems, including the absence of an adequate regulatory framework for high-risk investing as well as a lack of expertise in assessing credit risk or market opportunities. Although there is ample capital in private hands in China today, very little of it is invested in venture capital funds because of lack of confidence in the institutional environment. However, this has not stopped local governments, universities, state-owned companies, and foreign institutions from rushing to set up venture capital funds in China.

By 2000 there were approximately 160 domestic venture capital firms in operation in China, primarily located in Beijing, Shanghai, and Shenzhen. The government was the primary source of capital for most of these funds, either directly or indirectly through university- or state-owned firms. This seriously compromised the incentive for fund managers to make high-risk investments in private enterprises.[70] An investment in a government-owned company, by contrast, carries very little risk. Julie Yu Li, a partner in a venture capital firm started by a government-owned trading company in Shenzhen, summarizes the challenge: "I am supposed to invest in high-technology businesses, but my director once asked me if I could 'reduce the risk to zero'!" Fund managers like Li typically have little technical or business understanding of the industries they are investing in; they lack procedures for evaluating potential projects and ideas; and they have limited understanding of management and corporate governance. As Li points out, "We have no way to identify entrepreneurs or to evaluate risks and returns, and we must get approval from the president of the company to make any investments, which takes forever because it's very hard in China to get access to senior management."[71] A Silicon Valley–based entrepreneur who has advised the Chinese government on reform of the industry noted: "Venture capital fund managers in China have little at stake in the success of their ventures. If they are honest, they take no risk at all; if not, they take advantage of the opportunity to make under-the-table deals with entrepreneurs."[72]

The regulatory framework for venture capital investing is weak—it is unnecessarily restrictive in some ways, overly lax in others, and the rules are often overlooked in practice.[73] China's Company Law, for example, historically limited the number of shareholders and set a minimum level of investment that was not only high in the Chinese context but also exceeded the practice in the typical western VC firm. The law also limited the contribution of intangible technology to the valuation of an enterprise to 20 percent—which is unrealistic for technology businesses.

Reform of the system has been continuous in recent years, but remains uneven. In 2001 the government issued a notice allowing foreign companies to set up wholly owned or Sino-foreign cooperative venture capital firms, but they were not allowed to invest in securities, futures, or other financial markets; nor were they permitted to make loans or to underwrite or invest with borrowed money.[74] Another regulatory change recognized the limited liability partnership structure that is common in the West. However, the failure to specify qualifications for general partners makes it likely that many more venture capital firms in China will fail.

The venture capital industry faces other challenges related to the limited government oversight, including the information asymmetries between investors and fund managers, and between venture capital firms and the companies they finance. In both cases the lack of transparency, reporting standards, performance measures, and external oversight creates strong incentives for corruption and for concealing or falsifying information.[75] It is therefore no surprise that while China's venture capital firms have financed thousands of high-tech enterprises, these investments have generated minimal returns to date.[76]

Along with its banks, China's capital markets are the weakest link in its economy. Unlike the Taiwan stock exchange, which played a vital role in providing liquidity for venture-capital-financed start-ups, the Chinese stock market was organized in the 1980s to provide capital for the expansion of state-controlled companies. This bias against private enterprise continued throughout the 1990s as the Chinese exchanges were used to prop up failing state-owned enterprises. Even today, despite many efforts at reform, the market in China remains poorly regulated and subject to price manipulation. The old governmental quota system, which gave each province an annual IPO quota and ensured that state companies dominated the market listings regardless of their profitability, is being phased out.

However, the top regulators at the China Securities and Regulatory Commission are government appointees, and management of the exchange remains highly political.

The onerous listing requirements and regulatory procedures for approving public share offers on the main Chinese stock exchanges still favor state-backed firms. It took four years of intense lobbying for UFSoft to be listed on Shanghai's main board, in spite of its market leadership in accounting software, because of Chinese regulators' reluctance to approve the listing of privately owned firms. As of 2000, only a handful of the more than 1,000 companies listed on the Shanghai and Shenzhen stock exchanges were from the private sector.

In the late 1990s the China Securities and Regulatory Commission (CSRC) planned a second trading board in Shenzhen, modeled after NASDAQ, for high-risk, high-return companies. However, the anticipated opening was delayed following the U.S. technology stock collapse in early 2001. The CSRC officially approved the Small and Medium Enterprise Board (SME Board) in 2004 as a section of the Shenzhen Stock Exchange, with lower listing requirements for smaller technology companies. However, in the first eleven months after the launch only forty companies had listed on the SME Board. When a local government issues a list of the industries that venture capital will focus on in the coming years, this suggests that China is still trying to "grow the new economy using the tools of the planned economy."[77]

Relationship Marketing

Even if a private enterprise in China has access to capital and skill, it faces the daunting challenge of penetrating markets that often remain dominated by the state or by state-owned enterprises. One Chinese returnee describes the intimate ties between the government and the state-owned enterprises as "one body with two different heads," suggesting that it is misleading to view them as separate entities. *The Economist* concludes that a company's best defense against corruption and the political links that benefit competitors in China is to avoid collaboration (with customers, suppliers, or trade associations) or long-term technology investments, and instead build connections with the Communist Party hierarchy and curry favor with local bureaucrats. Scholars of China's economy have reached similar conclusions.[78]

Experienced managers acknowledge that the key to business success in China lies in "relationship marketing." The most successful Chinese technology companies, like Huawei and Lenovo (Legend), gained an advantage from strong early connections to the government. These firms have benefited as well from the Chinese leadership's preference for domestic technology products for security and nationalistic reasons. The desire to create strong domestic technological capabilities has led policymakers to invest heavily in Chinese businesses—even those producing inferior copies of Western products—in order to grow indigenous competitors. In some cases they have imposed rigid fiats requiring departments or state-owned enterprises to purchase Chinese-made components or services.[79]

The private technology enterprises in Zhongguancun and Zhangjiang science parks are typically started by graduates of China's elite universities or employees of research institutes. These entrepreneurs may be technically talented, but they rarely have the knowledge of markets or the connections to customers needed to commercialize their products. This challenge is particularly acute for technology enterprises because the government dominates in the market in both telecommunications and finance, either directly or through the state-owned enterprises. A Silicon Valley returnee who started an enterprise software company in Beijing puts it bluntly: "If you want to make big sales in China you have to be political. The telecommunications and banking sectors make up 50 percent of the

IT budget, and their structures still reflect the bureaucratic hierarchies of the planned economy. This means you are always dealing with the government."[80] In the financial and telecommunications sectors, the regulatory function is not separated from either the government or the operator, which creates conflicts of interest and uncertainty.

The government also controls technology markets in other, less visible ways. For example, all software products must receive evaluation and certification by the Ministry of Finance before they can be marketed in China. "Market entering permission" is also required before a product can be sold in a local market. So, for example, only three types of accounting software might be permitted in the Beijing market even though some fifty financial software products are available in China. Once again, government connections are a tremendous asset in this process.

Even apparent exceptions confirm the importance of government connections. UFSoft is a private enterprise that was started in Beijing in 1988 and is now China's leading producer of accounting software. The founders of the firm were former employees of the Government Administration Office's Finance Department: one oversaw development of the governmental accounting system and finance-related computer applications for State Council–affiliated institutions; another was in charge of developing the standardized accounting forms that later provided a template for UFSoft products; a third was in charge of enterprise accounting information, which involved analysis of data from 38,000 state-owned enterprises and 20,000 collectively owned firms. He explains that "the government experience helped us gain users' trust quickly. We already had relationships with many of the potentially large customers [governments and state-owned enterprises] and this made it much easier for us to market our products, especially in the early stages."[81]

UFSoft's "top to-bottom" marketing strategy likewise reflected the knowledge of its managers: the company markets products first to the central government, where it already has close relationships; this makes it significantly easier to sell to local governments. A former UFSoft marketing director reports: "It is much easier to market software to local governments than the central government because usually only one person is in charge, [whereas] in the central government you have to convince many different people."[82] About 95 percent of central government departments in Beijing now use UFSoft products, and the firm has 500 to 600 agents and fifty or sixty subsidiary companies throughout China, most of them focused on building relationships with local governments.

It is not surprising that successful firms in China today are drawing on the knowledge, expertise, and relationships that once resided in the state and the state-owned enterprises. Individuals or firms with access to the market knowledge, relationships, and capabilities (as well as the capital) from the state-owned era are best positioned to succeed in the new competitive environment. It seems less likely that domestic private enterprises that lack such resources—or outside resources—will be able to succeed in China today.

Private technology enterprises in China are also constrained by the lack of specialized business and professional service providers such as lawyers and accounting, marketing, insurance, public relations, and consulting firms. This need is particularly acute in an environment characterized by great regulatory uncertainty and complexity, corruption, and a poorly functioning legal system. The professional service providers in the Zhangjiang High-Tech Park, for example, are oriented to serving either multinational or very large domestic companies, which is reflected in their prices. This can create other problems. Since aspiring entrepreneurs in China rarely know how to develop a viable business model, they often adopt the traditional Chinese family-firm model, with husband and wife running the business along with other family members; this often leads to struggles for ownership and control of the business.

The isolation of China's small technology enterprises is compounded by the limited development of forums for informal information exchange in China, where organizations of civil society have been outlawed or discouraged for decades. Although a few business associations have emerged over the past decade, such as the China Semiconductor Industry Association and the Beijing Software Industry Association, they are closely tied to the government and their main role is to either lobby for or carry out policy—rather than providing the kinds of informal support and role models for start-ups that the professional and technical associations in Silicon Valley offer.[83]

A scholarly analysis of Zhongguancun Park in the late 1990s concludes that the weaknesses of its entrepreneurial firms resulted from the "strong hierarchical restraints from state-owned institutions or firms on local networking."[84] Even small foreign firms complain about the numerous intangible obstacles to development: "Foreign investors, especially small and medium-sized businesses, will face a maze of sometimes conflicting or unclear rules and regulations and an immature financial system."[85] Zhongguancun and Zhangjiang are also becoming overcrowded and expensive as both foreign and domestic firms seek access to potential customers and preferential policies: office space in the parks is increasingly costly, congestion in the surrounding areas is severe, and the costs of technical and managerial skill are rising fast. These costs further disadvantage private technology start-ups, particularly those that lack government connections or the deep pockets of foreign corporations or investors.

The Shanghai Model

The Zhangjiang Park Administration (and the Pudong New Area government) promoted technology development in the 1990s by courting foreign investment rather than by seeking to grow technology innovation and entrepreneurship. This was a natural strategy for the Shanghai bureaucracy, which established the city as a national center of heavy manufacturing (automobiles, electrical machinery, and the like) in the 1960s and 1970s by financing large state-owned enterprises and groups in exchange for managerial oversight through representation on boards.[86] The city developed a strong reputation for its predictable (if extensive) regulatory structures, its relatively efficient and transparent bureaucracy, and the least corruption of any large Chinese city.

Despite these factors—the city's good reputation and historically strong connection to global markets, as well as the largest and wealthiest concentration of consumers in the domestic economy—the process of attracting foreign investment to the Zhangjiang Park was slow at first. The first U.S. investor in the park, Allied Signal, built a turbocharger manufacturing plant in 1993 that would no longer be allowed today because it would not qualify as high-tech. After attracting a second foreign tenant, Hoffman Roche Pharmaceutical Corporation, Zhangjiang's planners began targeting more R&D-intensive activities in the mid-1990s.

The turning point for the park was in 2000, when the Shanghai government, accustomed to mobilizing resources selectively to support very large-scale enterprises, issued the "Stipulation for the Development of Shanghai Zhangjiang High-Tech Park." This policy focused government resources for high-tech industries on Zhangjiang rather than spreading them among the six existing municipal high-tech parks, as it had done in the past. Adam Segal describes how the Shanghai Tenth Five-Year Plan (2001–2005) identified IT as the leading sector in the city and called for local resources to be channeled into six domestic industrial groups that could compete both at home and overseas. This allowed the park planners to invest heavily in the physical infrastructure of Zhangjiang, now among the most advanced and modern in the world.[87] Large areas of the park are allocated to high-tech research, incubation, and manufacturing; there are also ample residential, commercial, and educational facilities, green space (40 percent), and efficient mass transit links.

The park administration aggressively courted technology investments with the provisions of Shanghai Circular 54, "Some Policy Guidelines of the Municipality for Encouraging the Development of the Software Industry and Integrated Circuit Industry," which provided loan subsidies, reduced fees for land use, utilities, and infrastructure, streamlined customs procedures, and rebates or exemptions on local and provincial taxes. Foreign enterprises were also exempted from local income, land use, and housing taxes for the first three years after their establishment.[88] The city government invested heavily in telecommunications as well: Shanghai was the first Chinese city to use ISDN (integrated services digital network) broadband and pioneered the deployment of high-speed, affordable ADSL (asymmetric digital subscriber line) connections for workplaces and residences. This strategy paid off in 2000 when SMIC, GSMC, and the domestic producer Shanghai Bell all committed to locating large IC manufacturing facilities in the park, accounting for over $3.4 billion (in U.S. dollars) in foreign investment—triple the average of previous years.

The roster of tenants of the Zhangjiang High-Tech Park includes 192 start-ups, and the park advertises a "high-tech innovation service center" to help start-ups find financing, provide operational assistance, and organize activities such as lectures. Yet few of the private entrepreneurial firms in the park are commercially successful. Hua Ming, executive director of Zhangjiang, believes this is because Shanghai has oriented its policy and institutions toward very large-scale enterprises; in his words, "People who work for large corporations or multinationals don't like to take the risk to start a company."[89] Shanghai's extensive government oversight of the Internet, as well as the technology arena, also appears to discourage private capital and entrepreneurship. This contrasts starkly with the situation in Taiwan's Hsinchu Science Park in the late 1980s, where everyone wanted to be "the head of an ox rather than the tail of a chicken," and where local social networks, government policy, and capital markets, as well as ties to Silicon Valley, supported a flourishing small- and medium-sized firm sector and a multiplicity of technology start-ups.

CROSS-PACIFIC ENTREPRENEURSHIP

The most successful returnee technology ventures in China today are "cross-Pacific start-ups" that combine the distinctive strengths of Silicon Valley and Greater China (both Taiwan and mainland China). The post-2001 returnees to China have apparently learned from their predecessors and, while still facing formidable challenges, have developed more sustainable business models. Their experiences suggest that the best business opportunities in China today involve applying established technology to serve the distinctive needs of the domestic market, and perhaps the markets of other developing economies, rather than seeking to advance the technological frontier or compete with established producers in wealthy markets.

The sustainable cross-Pacific start-up, according to Ronald Chwang of Acer Technology Ventures (the investment arm of Acer Computer) is a technology business that combines Silicon Valley's new product and business vision, technology architecture, product marketing, and R&D coordination with Greater China's R&D implementation, manufacturing and production logistics, and field engineering and local sales support. This combination leverages the management experience of Silicon Valley's overseas Chinese, China's low-cost engineering resources, standard and commercially available development tools, and the sophisticated supply-chain manufacturing infrastructure of Taiwan. In this vision, the U.S. market can serve as an early technology driver for new products like mobile appliances or new semiconductor designs, which can then be commercialized for both China (a low-cost version) and the United States.

These cross-regional start-ups nonetheless face substantial challenges. Venture investment in China is still in its early stages: there has not yet been a complete cycle of investments and reinvestments in a second generation of entrepreneurs. Cross-regional firms must overcome the substantial problems of coordinating distant activities, particularly in developing organizational synergy and ensuring persistent and consistent communication. Finally, like all technology firms in China, these start-ups face the challenge of controlling and protecting their intellectual property.[90]

The most successful of these ventures have adopted existing technologies to new products tailored to the distinctive demands of the Chinese market, including telecommunications equipment (UTStarcom), components (Comlent), and services (AsiaInfo). Another less visible group is leveraging China's low-cost engineering talent to serve export markets. These firms include companies that manufacture components (Sinoceramics) and design advanced ICs (New Wave), as well as a new generation of companies providing system integration or engineering outsourcing services (SRS2, Achievo). However, the rapid success of ViMicro, a state-sponsored, privately controlled IC design business, suggests that cross-regional firms will likely face stiff competition from technology entrepreneurship "with Chinese characteristics." Although the software and semiconductor industries are not dominated by the traditional state-owned enterprises found in China's older industries, the top companies in these technology sectors are often indirectly owned, run, and/or controlled by government officials in spite of their official status as "private" or "collective" enterprise.

Cross-Regional Investors

The only authentic high-risk venture investors in China are some fifty foreign venture capital firms, including several veterans from Silicon Valley and Taiwan. These firms—including Japan's Softbank, Warburg Pincus, Intel Venture Capital, WI Harper, Walden International Investment Group, H&Q Asia Pacific, Acer Technology Ventures, V2V, IDG, and Vertex (Singapore)—have invested in establishing full-time management teams in China. They invest cautiously in China, however, because of the lack of viable exit options. Access to China's SOE-dominated stock market is close to impossible, and the non-convertibility of the renminbi (RMB), the official currency of China, makes it difficult for foreign investors to repatriate earnings.

A series of high-profile IPOs including SMIC, 51job Inc. (an online human resource services company), and China's largest online game company, Shanda Interactive Entertainment, fueled a rush of new investors into the China market in 2004, boosting venture capital investment to just over $1 billion, or double the amount a year earlier. While this is not a large amount compared to the total foreign direct investment in China of approximately $62 billion in 2004, many observers worry about a "mini-bubble," with too much risk capital chasing too few high-quality start-ups alongside already excessive valuations of Chinese technology stocks. David Chao, an overseas Chinese venture capitalist who is based in Silicon Valley but travels to China every six weeks, observes: "I think of China as the wild, wild West. Business law has only been in practice for the last 10 years. There are fundamental issues that still have to be addressed. What rights does a shareholder have in China? There's also this significant cultural gap. . . . There's definitely a lemming effect going on here. A lot of these guys will end up getting burned."[91]

Longer-term investors have learned from hard experience the challenges of doing business in China. The chairman of Walden International Investment Group (WIIG), Lip-Bu Tan, recalls that when he first went to China in the early 1990s he intended to invest in state-owned enterprises, but soon realized how difficult it would be to overcome their inefficient management, obsolete equipment, and limited understanding of market economies. He then sought to create joint ventures between U.S. start-ups and local Chinese companies, but concluded that the challenge of bridging the conflicting goals and approaches of two such different management cultures was insurmountable. According to Tan, the experience with Howard Yang's New Wave Silicon taught him that the best strategy in China is to invest in U.S.-educated graduates who want to return home to start firms.

Newave Semiconductor was acquired in 2002 by Integrated Device Technology, a Silicon Valley–based semiconductor company, for $85 million, earning profits (and U.S. dollars) for both WIIG and China's Hua Hong Microelectronics. The key, says Tan, is to find graduates of U.S. universities who have stayed and worked in companies in a place like Silicon Valley for many years: "You have to be reasonably brainwashed in the U.S. to succeed in a western-style technology venture in China." As the mainland Chinese community in the United States continues to mature in the first decade of the twenty-first century, such seasoned start-ups seem increasingly plausible. Tan's vision of the model returnee firm, like that of Acer's Chwang, includes headquarters and R&D capability in Silicon Valley, incorporation in the Caribbean, manufacturing and/or development in China, and higher-level design or logistics capability in Taiwan, led by entrepreneurs who understand Chinese language, culture, and institutions.

The *Xiaolingtong* Supply Chain

UTStarcom is one of China's most successful returnee businesses, although it took many years for the firm to find a niche. UTStarcom's

Chinese operation started as a small telephone switching equipment factory in Hangzhou in the early 1990s. A decade later, the firm went public on NASDAQ as the leading provider of wireless and wired access and IP switching products and services in China, and the world's largest supplier of personal access system (PAS) equipment and services. PAS is called *xiaolingtong* in China, translated as "little smart one" or "smart little connector," because it provides wireless access services within a certain geographic area at a significantly lower cost than the cell phones developed for the Western market. The PAS technology was developed in Japan, but it never took off there because the network quality is relatively poor and doesn't allow roaming between cities. However, the service is cheap—about half the cost of standard cell phone service—which appealed to millions of lower- and middle-income Chinese. The major telephone companies, China Netcom and China Telecom, also liked the system because it allowed them to offer a form of mobile service even though they were not permitted to offer cell phone service.[92]

UTStarcom capitalized on the fast-growing demand for mobile phones in China (PAS service reached 70 million units in 2004, compared to 1.5 million in 2001) and has taken advantage of the efficient, low-cost Taiwanese IT manufacturing infrastructure in the Hangzhou area to mass-produce its handsets. The company's rapid growth has also encouraged other returnee start-ups to develop innovative technology inputs (Comlent in radio frequency ICs and Jazz Semiconductor in the silicon germanium foundry process) for this successful product.

The original company, Starcom Networks, was founded in 1991 by a team of U.S.-educated mainland Chinese executives who had worked together at Bell Labs in New Jersey. UTStarcom is the result of the merger in 1995 of Unitech Telecom, a California developer of digital and wireless transmission systems, with Starcom Networks. Although UTStarcom is incorporated in Delaware with its headquarters in Alameda, California, the firm set up a factory in Hangzhou, China—the hometown of founder Hong Lu—almost immediately. In 2004 UTStarcom's R&D center in Silicon Valley employed some 1,000 workers, but the majority (over 80 percent) of its 6,500 employees were based either in the company's R&D and production centers in Zhejiang province or elsewhere in China doing marketing, sales, and services. The firm recruits engineers from a nearby university run by China's Ministry of Posts and Telecommunications, many of whom are sent to the United States for eighteen months of training to ensure that they meet the company's standards.

Co-founder and CEO Hong Lu emphasizes that UTStarcom has profited from its connections in China; the advantages include knowledge of the China market and the ability to combine modern business structures and competitive products from the United States with the founders' ethnic and cultural know-how.[93] He believes that having mainland Chinese founders has helped the firm compete successfully in China with multinational telecom companies like Lucent and Motorola, as well as helping convince Chinese officials that the firm was in the market for the long haul. Their mainland Chinese origins also ensure that they know "how to function in the local culture and how to contact phone companies."[94] Early subscribers to UTStarcom's *xiaolingtong* service were gained largely by building close relations with provincial officials, including the establishment of joint manufacturing ventures with Posts and Telecommunications Administration officials in Guangdong and Zhejiang provinces.

UTStarcom depended as well on access to ample and patient funding from Masayoshi Son, the high-profile head of Japan's Softbank Corporation and a personal friend of founder Hong Lu. Son and Lu met in the early 1980s when Lu ran an ice cream parlor in Berkeley to help support his graduate studies. Softbank invested some $160 million in UTStarcom during the 1990s, which gave the company several years to find a viable niche in the Chinese market. The firm was publicly listed on NASDAQ in 2000. In February 2001, a decade after the firm was started, UTStarcom had only a million subscribers. By 2005 the firm had 47.5 million PAS subscribers, or 60 percent of all Chinese subscribers—most of whom were urban workers in low-paying service industries such as street vendors, garbage collectors, and waiters and waitresses who needed phones but couldn't afford more costly cell phones.[95]

UTStarcom operates a state-of-the-art final assembly and test facility in Hangzhou, but the firm subcontracts its high-volume manufacturing and assembly of its circuit boards and handsets to local vendors. Taiwan's Foxconn Electronics, one of the firm's lead suppliers, expanded its Hangzhou facility to twenty surface-mount technology production lines in order to deliver 16 million handsets to UTStarcom in 2003. Foxconn also set up a precision mold and die development subsidiary in nearby Kunshan. Although it took more than a decade for UTStarcom to grow to its current size, and there are few comparable entrepreneurial growth stories in China, the flocking of domestic imitators was rapid. Dozens of "me-too" manufacturers of *xiaolingtong* handsets emerged, ensuring that the price remained low in an intensely competitive market.

The overseas Chinese community has utilized technology, product development experience, and financial resources from Silicon Valley and Taiwan to serve promising niches in China's fast-growing wireless communications market. Comlent Holdings was started in 2002 by a group including a number of returnee engineers to design the radio frequency chips (RFICs) that are used in PAS handsets like those sold by UTStarcom. The RFIC market was ignored by mainstream U.S. and European IC firms because the devices are used primarily in China and Japan. Comlent is headquartered in Shanghai, with an R&D center in California and a back-end production team in Taiwan. Comlent co-founder and CEO Kai Chen, who earned a Ph.D. at U.C. Berkeley and worked in the U.S. semiconductor industry for fifteen years, explains that the firm differentiates itself from the hundreds of copycat IC design firms in China by integrating as many as six radio frequency chips into one or two chips: "We're not just using Chinese labor to make the chip 10 percent cheaper; we're revolutionizing the whole architecture."[96]

Chen believes that Comlent benefits from proximity to its customers, the mainland handset manufacturers, and its knowledge of their customer base (the end-users). The firm's chipset is designed to address the frequent complaint among China's PAS users that their calls are dropped; the higher level of integration not only means that the IC should perform better but also will reduce the cost of a handset by 10 percent or more. This strategy of customizing and optimizing IC designs for local demand has been used effectively by Taiwan's IC design industry over the past decade. According to Chen, most American IC firms design chips and expect their customers to develop products that use them. He adds: "The Chinese manufacturers try to build systems around that chip—good or bad."[97]

Comlent has benefited not only from the U.S. education and technical experience of its founding team, but also from the entrepreneurial experience and connections of its initial investors. Ming-Kai Tsai, co-founder and chairman of Taiwan's MediaTek—the leading designer of chips for DVD players—and seasoned Taiwan–Silicon Valley investor Herbert Chang were first-round investors in Comlent. They were undoubtedly helpful in arranging a foundry partnership for volume production with Taiwan Semiconductor Manufacturing Company (TSMC). Comlent raised $7 million in its second round of financing from Silicon Valley–based investors, including Intel Capital and Draper Fisher Jurvetson.

Jazz Semiconductor, Comlent's other foundry partner, is a California-based start-up (2002) with a U.S.-educated, mainland China-born CEO, Shu Li. The firm, which spun out from Conexant Systems, specializes in an emerging silicon germanium (SiGe) foundry process geared to low-cost integrated wireless and wire line applications. This focus on a very-high-speed SiGe process differentiates Jazz from the large, established foundries such as TSMC and UMC, with their advanced digital CMOS process technology capabilities. Although the first Jazz foundry is based in the United States, the firm's senior management has announced that future expansion of manufacturing will be located in China.

China will then be home to a complete PAS supply chain—from IC design and manufacturing to handset production and services. UTStarcom is well positioned to collaborate with experienced radio frequency chip designers like Comlent to improve the quality and functionality of the product while also lowering costs, and eventually to serve not only China's wireless market but also those of other developing countries.

Of course the China market is not for the weak-of-heart, and competition in the telecommunications market is intense. In 2005, after five years of sustained growth, UTStarcom faced substantial losses as demand for its PAS business fell precipitously as a result of market saturation, competition at the low end from local equipment makers, and a transition among China's major telecommunications providers. Meanwhile, multinationals such as Nortel, Siemens, and Nokia are aggressive competitors in the China market. UTStarcom has undertaken a significant restructuring and seeks to expand its business into new markets such as traditional and next-generation cell phone services, high-speed Internet services, and IPTV (cable television service over high-speed Internet lines), while also actively marketing its products outside of China, in India, Brazil, and parts of Africa as well as the United States and Europe. Founder and CEO Hong Lu calls these changes "the new unknown."[98]

Innovating in IC Design

The cross-Pacific venture (start-up) model is gaining momentum in the IC industry, where mainland Chinese returnees are exploiting ever more specialized niche opportunities to advance semiconductor design technology. These cross-Pacific start-ups stand out among the hundreds of very small and frequently undercapitalized IC design firms that have proliferated in China since 2000.

VeriSilicon Holding was founded in 2001 by Wayne Wei-Ming Dai, who holds a Ph.D. from U.C. Berkeley, with the goal of becoming a leading ASIC "design foundry." The firm offers a standard design platform optimized for the leading foundries in China as well as a collection of both internally developed and acquired semiconductor intellectual property (SIP) components that it uses for both its ASIC (system on chip) design services and ASIC turnkey services. With its headquarters in Shanghai and operations in both Silicon Valley and Taiwan, VeriSilicon is positioned to tap the capabilities of the world's three leading IC centers. Dai, who started two companies in Silicon Valley prior to VeriSilicon (including Celestry Design Technologies, which was acquired by Cadence Design Systems in 2003), has recruited senior executives with U.S. graduate education or work experience in Silicon Valley, or both. His vision is to combine American strength in innovative designs, architectures, and equipment with Taiwan's foundry expertise and China's low-cost engineering talent and fast-growing market.

The cross-Pacific venture capital firm Authosis specializes in early-stage fabless IC design companies, particularly those focusing on the China market. The firm is headquartered in Hong Kong and has operations in Shanghai and Shenzhen as well as in Silicon Valley. Authosis founder and CEO Danny Liu has extensive experience working for China's Legend Group (now Lenovo), which provides invaluable relationships in domestic technology and business circles, as well as experience investing in Silicon Valley start-ups. The advisory board of Authosis includes the former CEO of the Silicon Valley Bank, a former dean of engineering at U.C. Berkeley, the former CTO of TSMC (also a professor at U.C. Berkeley), and computer science and engineering professors from MIT, Tsinghua University, and Peking University. One of the firm's earliest and most successful investments was the electronic design automation company Celestry Design Technologies, started by Wayne Wei-Ming Dai.

Entrepreneurship with Chinese Characteristics

The Chinese government, recognizing the opportunities created by the changing division of labor in the IC industry, has targeted investments in IC design as well. The returnee start-up Vimicro Corporation, for example, gained national attention in 2004 for its large share (43 percent) of the world market for digital imaging processors for PC cameras. Vimicro received funding from the powerful Ministry of Information Industry, which does not control or manage the firm, but the MII has provided introductions to key customers and promoted the firm's technology as a national standard. This distinguishes it from the hundreds of other undercapitalized IC design firms in China. In 2005, Vimicro's "China Chip Project" was praised by the Party Congress as a historical national achievement (along with the space flight and gene sequencing).

The Chinese government sponsored the first national semiconductor design center in Shanghai in 2000, and soon selected six other cities to become chip design bases, often located near a major electronics manufacturing plant. These bases serve as incubators for IC design enterprises and provide a range of technical training and consulting services, as well as low-rent office space. However, repeating the pattern found in the science parks, most of China's fabless IC design firms remain two- or three-person operations run by recent engineering graduates who lack expertise in application design, marketing, management, or sales. Many are imitators that reverse-engineer designs and compete solely on the basis of cost.

A survey by iSuppli Corp found that in 2003 the majority of China's approximately 486 fabless design companies were seriously undercapitalized. They collectively employed only 7,000 workers, of whom 60 percent had fewer than four years of experience. Chinese electronic systems manufacturers remained reluctant to use their services because of concerns regarding quality and availability—with the result that domestic fabless IC designers were serving only 1.5 percent of China's demand. Ten companies accounted for 72 percent of sales, and only a dozen of these design firms generated sales exceeding $12 million. They were also behind technologically: the majority were designing for older-generation process technologies of 0.25 or 0.35-micron or greater. Only 20 percent of these companies could complete mainstream 0.18-micron designs, and very few (5 percent) could handle designing for leading-edge 90-nanometer processes. In other words, chip design in China remains at relatively low levels of integration, with low transistor counts and clock speeds, reflecting a focus on smaller analog, mixed signal, and low gate count digital ICs for consumer electronics, communications, and automotive applications.[99]

Analysts predict that IC design production will grow rapidly in China as a result of aggressive government support and the growing domestic market—and in spite of the lack of adequate skills. Chinese universities were producing only 400 IC design engineers annually in the early 2000s, yet domestic demand for these IC designers was forecast to exceed 20,000 in 2006. As a result, China's fabless chip design firms, along with their manufacturing counterparts, are likely in the near future to remain suppliers of low-cost, trailing-technology chips for consumer electronics products such as cell phones, video game devices, desktop and laptop computers, and televisions. In the longer run, the expansion of China's higher education system combined with improved R&D capability and a deeper experience base will reduce their dependence on imported IC designs.

CHINA'S SOFTWARE BOOM

Although China's semiconductor and telecommunications-related companies are the most visible evidence of growing ties to Silicon Valley, the overseas Chinese community has pioneered a small but fast-growing software and engineering outsourcing business. This effort is far behind that of India, where software services firms have decades of experience as well as the advantage of a large English-speaking skill base, the ability to conduct business with Western clients in their cultural idiom, an established pipeline of clients, and an international reputation for reliability and credibility. China's software revenues grew more than 40 percent per annum between 1997 and 2003, when they reached an estimated $6.8 billion, but still remained half of India's $12.7 billion.[100]

India's head start has not, however, stopped Chinese entrepreneurs from drawing on their language skills and contacts to attract U.S. customers to the large, low-cost software and engineering skill base in China. They have leveraged the sophisticated physical and services infrastructure and the environment conducive to foreign managers that has developed over two decades of IT manufacturing outsourcing. The number of English-speaking graduates in the workforce (which is key for outsourcing) has doubled since 2000, to more than 24 million in 2004. Growth has been so rapid that some industry observers predict that China will achieve revenues in software services comparable to India's before the end of the decade.[101] In the words of a leading Silicon Valley entrepreneur:

> In the mid-'90s, if you wanted to design a new protocol for the Internet, something that would become the eventual standard, this

[Silicon Valley] was the place to recruit the software engineers to do that. Today a networking equipment vendor in China, which we have a very significant joint venture with, has people who know as much about the Internet as anyone else who attends the IETF [Internet Engineering Task Force] meetings. Similarly, if today you want to build a 50,000-gate ASIC, chances are you could build this with an ASIC lab over there as well.[102]

Achievo and Its Followers

Robert P. Lee started one of the first U.S.-based Chinese outsourcing companies, Achievo, in 2002. Lee is an old-timer in the Bay Area computer industry: a Hong Kong native, he was among the first generation of Chinese engineers to come to Silicon Valley during the 1970s. After graduating from U.C. Berkeley and UCLA, he worked at IBM and Symantec, and he was CEO of Inxight Software and of Insignia Solutions (one of the first Chinese-run technology companies to go public on NASDAQ). Lee raised funds for Achievo from the investor Sandy Wai-Yan Chau, also a Berkeley graduate who has worked in Silicon Valley since the 1970s.

Achievo is small by global standards, with approximately $10 million in revenues and three hundred employees in 2004, but it has grown fast. The company promotes itself on the basis of a U.S.-trained management team based in both Silicon Valley and China that ensures the language and cultural compatibility and the accountability needed for continuous communications. Raymond Tong, vice president of Achievo's China operations and another Berkeley graduate with Silicon Valley experience, is explicit: "Our mission is to become like Infosys and the other big Indian companies."[103]

Tong runs Achievo's China operations from Shenzhen, which provides easy access to Hong Kong, to low-cost programming skill (salaries are roughly a sixth of those in the United States and comparable to India's), as well as to the tax breaks, low-cost space in science parks, and other incentives provided by the Chinese government for the software industry. The firm also has a branch in Beijing. Achievo's Web site promotes its business in part by stressing the challenges that India, with its limited supply of IT graduates and experienced project managers as well as its infrastructure problems, will face in meeting the fast-growing demand.

San Francisco–based Silk Road Software Services (SRS2) is also leveraging China's low-cost technical skill with U.S.-educated management talent. SRS2, which employs more than three hundred workers in development centers in Shanghai and Beijing, describes itself as an IT outsourcing and software engineering services provider. Larry Chen, the president and co-CEO, explains the firm's vision of moving beyond "simple coding" to providing enterprise resource planning, logistics, and supply chain management services to link global activities of large U.S. firms with their activities in China.

Chen uses the example of San Jose–based WebEx Communications. The founder and CEO of WebEx, Min Zhu, came to the United States from China in 1984 to earn a graduate degree at Stanford University. After working at IBM for several years, and then starting and selling a company to Quarterdeck, he founded WebEx with an Indian colleague. Unlike most other mainland Chinese entrepreneurs in Silicon Valley at the time, Min Zhu did not seek to serve the China market: "I am a Silicon Valley person. Silicon Valley gave me a unique opportunity, and I am focused on being a leader in technology and business, not on connections in China."[104] Following the successful WebEx IPO in 2000, however, he began exploiting his connections in China to outsource the firm's software development and testing activities. Initially Min Zhu and his wife personally purchased three companies in the Shanghai area and subcontracted software services for WebEx from these firms. In 2003 a Chinese subsidiary, WebEx China, acquired all three companies and hired most of their employees. This 480-person team began to do research and development, or what Min Zhu refers to as "low cost exploration," such as adapting their network services and applications to new operating systems, performing lead research for the sales division, and providing client side development and technical support in Asia—using WebEx conferencing tools for communication and collaboration. Although the company's core technology remains in Silicon Valley, the low cost of labor in China allows WebEx to afford the "large army" needed to carry out the labor-intensive work of developing, implementing, and quality assurance testing of successive product versions.[105]

Larry Chen of SRS2 is following the WebEx model by building the infrastructure for collaborative software outsourcing and engineering, including business process management and applications architecture design and development as well as systems integration, legacy-to-Web design and conversions, upgrades and customizations, and the testing of quality assurance, porting, and localization. Chen is targeting senior Chinese managers in large U.S. corporations. In keeping with the experience of the Indian community in the United States, he finds these managers to be far more accessible than non-Chinese managers. Over time, he believes that the biggest outsourcing opportunities will be with the largest domestic Chinese manufacturing enterprises, which need to streamline their business processes and logistics. However, Chen recognizes that the internal market for software services remains quite small: very few Chinese firms, including Chinese multinationals, are willing to spend money on IT. As a result, he is building the firm's scale and reputation from the United States in order to serve the China market in the future.[106]

The Indian Frontier

More than a dozen Indian software services companies have opened offices in China since 2000 in an effort to gain access to its large talent pool, and a foothold in the market as well. In 2002 China had some 200,000 IT professionals involved in software export services, with about 50,000 entering the business annually.[107] In the words of an Infosys vice president of outsourcing services, "The opportunities are huge. You have an untapped workforce and a very motivated economy."[108] However, the giant Indian software firms like Infosys lack the language and cultural connections, as well as the understanding of the business context, that would allow them to compete in a market that depends heavily on relationships and trust. Their successes building an outsourcing business in the United States leveraged the intermediary role of senior Indian managers in big corporations—but few have these connections in China.

India's leading software training businesses, National Institute of Information Technology and Aptech, have nonetheless set up more than a hundred training centers in China and teach more than 20,000 students a year. The Chinese government, recognizing the need for English-language skills, has also invested heavily in expanding language training programs. (The symbol-based Chinese language initially posed major challenges for electronic communication, and thus for IT outsourcing, with the United States. However, this problem has been largely addressed with the development of Unicode.)[109]

China still lags behind India in the outsourcing arena, but local firms are fast developing a diverse range of capabilities in software products, engineering, and services. Technology and know-how are advancing quickly as a result of the joint ventures as well as research labs and training programs set up by foreign investors. The Intel China Software Lab, for example, has more than ninety engineers working on leading-edge software for applications ranging from DSP/multimedia products and device drivers to signal integrity technology. Learning happens abroad as well. China's leading telecommunications equipment vendor, Huawei, has a software center in Bangalore that in 2002 employed some four hundred Indian programmers.

Software with Hard Choices

Although China's software market is growing very fast, it remains immature. Many of the largest domestic software developers are diversified IT firms, such as Founder and Lenovo. The telecommunications equipment manufacturers Huawei and Zhongxing Telecommunications, for example, are the two largest software vendors in China; the only focused software firm in the top ten, Neusoft, ranked tenth in 2003. Most industry participants agree that few domestic users (government, businesses, or individuals) fully appreciate the value of investing in or paying for software. This is related as much to organizational immaturity as to resource constraints, as described by a vice president of the Beijing-based software firm Digital Wangfujing: "The biggest problem for the software industry in China today is that the users have not matured enough to use or to understand software. The government and firms still don't have a well-defined governance structure which can tell them when and how to hire or fire employees. How can we expect them to know when they need software?"[110]

Most Chinese business and government customers, having recently made the transition from a planned economy, lack the experience, confidence, and authority to make thoughtful, independent purchase decisions. In fact, the work of many enterprises and even local government agencies is still primarily based on paper and/or hand labor, so it makes little sense to talk about higher levels of automation or integration. Even managers who recognize the need to invest in software have great difficulty in evaluating investments and their potential returns because Chinese enterprises rarely know their actual costs; many also operate in "information silos" enforced by a government determined to control information.

In contrast with the Indian software industry, which has grown up serving extremely sophisticated foreign corporations, the backwardness of the domestic market has made it difficult for Chinese software firms to develop new products or achieve scale. After the government, the largest domestic customers for software in China are the state-owned enterprises. Though subject to aggressive restructuring over the past decade, the SOEs still account for the majority of the software market outside of technology sectors. Yet it is rare for a Chinese SOE to depend on software-derived or software-supported functionality for its competitive advantage. For most, competition is based on price, distribution, and relationships.

There are, of course, exceptions. Some sophisticated small- and medium-sized Chinese companies have fully integrated enterprise resource planning programs—but most systems in China still do not require advanced skill levels. Others fail to see the value of software for their operations. One software consultant reports, "In today's China you can find $30 billion multinationals with virtually no IT functionality besides e-mail and a marketing Web site."[111] However, as Japanese and Korean firms increasingly turn to China for provision of low-cost software and engineering services, and as Taiwanese firms invest in logistics-related software development in Shanghai, the Chinese software market should expand.

Competition from the State

China's Tenth Five-Year Plan identifies software, along with semiconductors, as a critical or "pillar" industry that is essential to economic progress and national security. A government that remains obsessed with information control frequently invokes military and security concerns to justify the need for domestic software capabilities. A 2000 editorial in the *People's Liberation Army Daily* on "information colonialism" argued, for example, that China must develop its own software because "without information security there is no national security in economics, politics, or military affairs."[112]

Returnee software enterprises therefore face large state-supported domestic competitors. At the same time that state-owned enterprises in traditional industries are being downsized, new ones are being created. China National Computer Software & Technology Service Corporation (CS&S, or ChinaSoft) was established in 1990 as a subsidiary of the state-owned China Electronics Corporation. It is recognized as a "flagship" enterprise of the MII. Based in the Zhongguancun Science Park in Beijing, ChinaSoft has more than thirty holding and shareholding companies that specialize in software and information product development, system integration, information service, and software development outsourcing. The company is now China's largest software development and systems integration firm, with more than 2,000 employees.

ChinaSoft was the first company in China to be classified as a "Software Enterprise," which means that it fulfills the administrative guidelines established by the government to qualify for major government software projects. Most of ChinaSoft's customers are SOEs, and it is the only software company listed as one of 520 National Key Enterprises since 1999. The company was also the first to pass the national security-related Systems Integration Qualification in 2002, and the only systems integrator in China to have obtained the National Systems Integration First-Level Qualification in 2000.

The Chinese government supports software firms like ChinaSoft through various programs. In 2002, for example, the Ministry of Science and Technology sponsored a project called the "China Offshore Software Engineering Project" to help advance the internationalization of China's software industry; the Ministry supported a hundred domestic software firms in developing exports to Europe and the United States. The Ministry assigns experienced consulting companies to help prepare the firms for exporting; those firms that develop export capabilities and complete two offshore software projects that satisfy customers' requirements receive financing of 60,000

RMB from the government. In addition, the government organizes delegations of representatives from Chinese software firms and software industrial parks to international conferences in order to help them find clients.

Chinese firms have failed so far to achieve the level of quality and reputation associated with successful outsourcing in India. Returnee firms, however, have the potential to provide models of efficient organization and management of software development, particularly by demonstrating the use of incentives to attract and retain skilled employees. Government agencies are also beginning to provide financial incentives for software companies to engage in Carnegie Mellon's SEI capability maturity model (CMM) certification. The Shanghai Software Quality Consortium was established in 2000 to promote software process improvement.[113]

Buy Chinese

The government is the dominant paying customer of the Chinese software industry. Government ministries, agencies, and institutes at all levels procure software to support the goals of national industrial development as well as to improve their own productivity—and they are frequently urged to procure local rather than foreign-made software. Moreover, resources remain heavily concentrated in the state sector. Firms that have good relations with the government are well positioned to win contracts. Government purchasing often takes the form of large-scale projects such as the central government's Golden Projects for Informatization—Golden Card, Golden Bridge, Golden Custom, and so on, and the more recently instituted E-Government programs. In addition, a variety of city-sponsored Digital City and Information Port programs provide huge software development contracts in the large urban areas. As China prepares for the 2008 Olympic games in Beijing, the central government is developing a new generation of projects: the "Digital Olympics" project, for example, will provide extensive funding for software as well as other IT firms.

While the government is generally seen as a good client in terms of technical competence, ability to communicate, cooperativeness, and ability to pay, most Chinese government agencies (like Chinese businesses) are technically unsophisticated and unable to articulate their requirements or specifications clearly. Government procurement can benefit an emerging industry under certain circumstances, but this role is routinely compromised in China because of the government's other, often conflicting, roles. The strong desire among government officials, for either economic or security reasons, to create "national champions" leads them to invest heavily in domestic software companies, or to rigidly enforce purchasing fiats requiring that all departments and state-owned enterprises purchase Chinese-made components and services.

In 2001, for example, the Beijing municipal government required that all of its departments purchase WPS2000, a Chinese-language office software package developed by the domestic firm KingSoft, rather than Microsoft Office 2000—in spite of widespread agreement as to the technical inferiority of the product. (WPS2000 is far less stable than Office, and KingSoft provides virtually no after-sales service or customer support.) In 2002, following the lead of Beijing, the Guangdong provincial government purchased over 4,000 packages of WPS2002 software for use in more than forty departments and bureaus. In some cases, the government promotes domestic competitors by subsidizing firms like the Beijing University spin-off Jadebird (which continues to develop an indigenous database product to compete with Oracle and Sybase), or by providing contracts to start-ups developing Linux software products.

The administrative guidelines set by the Ministry of Information Industry, purportedly to ensure standardization and market regulation, also privilege state-owned enterprises and university spin-offs over their private competitors. All applicants for government and major SOE work are required to obtain a Certification of Capability and Quality, which also establishes a firm's eligibility for tax incentives and other promotional measures established for the industry. The standards for this certificate are based on traditional measures of scale like number of engineers, total net assets, registered capital, and annual revenues, but they have little to do with projects executed, work process, or quality. This makes it difficult for new private firms to meet the standards.

The dominance of the government market, as well as the absence of the rule of law, is reflected in the pervasiveness of "relationship marketing" in China. The importance of connections in obtaining contracts from Chinese enterprises and government institutions is widely recognized. Policymakers are attempting to change this, especially in the newer technology sectors, but software producers still spend far more time and energy building reputations and relationships than in the West. A Silicon Valley–based entrepreneur whose company provides content for the mobile Internet in China refers to this as the problem of "high transaction costs." He reports:

> For most of us doing business in China today, the biggest challenge is the time that we have to spend going to dinners and banquets, drinking, making friends, and building trust. The hard infrastructure here (like telecommunications and airports) works very well, but the soft infrastructure is still barely developed. If you can't depend on the legal system for remedies, you have to make sure that you trust your business partners, that you speak the same language, and know one another well. Since there are no rules, we have to do business on a case-by-case basis, which means building trust case by case.[114]

Foreign and private firms face particular challenges in this environment. A senior manager from Oracle argues that it is critical for a foreign company trying to compete in China to recruit salespeople with great care. The two criteria that Oracle uses in hiring for its sales and marketing team are the individual's capabilities and the quality of the individual's network of relationships. A strong relationship network, in this view, depends upon working experience either in related government sectors or for major customers in the target market, which themselves are likely to have strong government ties.

One indicator of the importance of personal connections in post-reform China is the experience of returnees from the United States. Many returnees overlook this factor, having worked in an environment where the quality of technology and service is the key to competitive success. Having spent more than a decade in the United States, most lack the personal connections needed to succeed in the Chinese business environment.

The pressures of the World Trade Organization are likely to continue to promote more open markets. China's WTO commitments required that after January 2003 wholly foreign-owned enterprises be allowed to enter the Chinese IT market, and that China phase out all tariffs in information technology. However, software sold in China must also meet the standards set by the Chinese Platform Standard Committee, which approves the sale of software in the domestic market; foreign software companies often need to modify their products to meet these local requirements. Some who do business in China report that the true value of the WTO is that it provides an excuse among reformers in the government to undertake far-reaching changes. Others believe that its main value is to provide the perception of competitiveness in China, even if there is a tendency among regulators to follow the letter rather than the spirit of their WTO commitments.

LOOKING FORWARD

Visitors to China are frequently impressed by the sophistication of government officials—many of whom speak fluent English, have traveled extensively abroad, are open to outside views, and are articulate about the challenges facing the corporate and financial sectors. Visitors who return regularly are also amazed by the pace of change at every level in Chinese society and the economy. This makes it difficult to look into the future with any certainty: China is changing very fast, and also very unevenly, and there are too many political as well as economic factors at play to predict much except for continuing change.

Returning engineers and entrepreneurs with links to their counterparts in Silicon Valley and Taiwan are contributing to the transformation of the Chinese economy by transferring technical, managerial, and organizational know-how as well as market connections. The government is investing aggressively in infrastructure and research while increasing market-based competition. Nonetheless, the state still controls the major economic assets in China. The institutional environment for commercial innovation and entrepreneurship is underdeveloped, and most private firms remain technologically and financially weak. As McKinsey consultant Jonathan Woetzel observes, "You can be a small entrepreneur in China, but if you want to be big, you will have to get money from a government-affiliated source at some point. Government officials essentially have the power to decide which companies grow."[115]

The circulation of engineering and entrepreneurial talent between the United States, Taiwan, and China is altering the economic trajectories of all three. China will continue to challenge the adaptive capacities of Taiwan's industrial system. While experts predict that it will take China a dozen years to reach technological parity in the manufacturing of most advanced ICs and IT products, Taiwan's producers are establishing new specializations, including new market niches, and developing improved logistics and distribution capabilities to further enhance their technological and operational advantages. Taiwan is thus developing a complementary relationship to

China as the former becomes a center for high-value-added design, leading-edge manufacturing services (including product implementation, logistics, and business operations), and marketing, while China focuses on product development and low-end, labor-intensive design and assembly-and-testing as well as non-leading-edge manufacturing. This is a dynamic situation that will be driven both by the growth of China's domestic market and by the falling tariffs expected to accompany China's integration into the WTO.

It is easy, however, to overstate both the scale of the Chinese market and its technological capabilities. Demand for technology products remains relatively low in China; the new fabs like SMIC, for example, serve primarily foreign customers and face continuing financial difficulties.[116] China's IC design, manufacturing, and software capabilities lag those in the United States considerably. Even the high-profile R&D labs of U.S. technology companies like Intel, Cisco, and Microsoft do not view Chinese companies as potential research or production partners, and their collaborations with local universities appear to be primarily a means to identify technical talent and to enhance their reputations and relationships with government officials.[117] China will likely continue to consume relatively low-end chips—the type used in watches, radios, cell phones, and other consumer electronics products—for the rest of the decade.

Although Chinese IT exports reached $210 billion in 2004, this commerce was overwhelmingly managed by foreign companies. Domestic Chinese producers lack marketing and managerial experience as well as connections to markets. It is one thing to learn mass manufacturing or to reverse-engineer a high-tech product, but quite another to develop competitive technologies and products. Even China's most successful technology companies remain followers; and the exodus of dozens of IBM-China managers following the Lenovo acquisition of the firm's PC unit suggests that it will be difficult to buy the capabilities needed for leadership.

Chinese companies and university researchers remain isolated from the market. Frank He, president of a Dallas-based software company, explains that "in China . . . we can barely find one out of five hundred Ph.D. dissertations (e.g., in computer science) with research findings useful for commercialization—while in the U.S., one dissertation out of one hundred can be turned into a real product."[118] China's research funding agencies such as the National Natural Science Foundation of China and the Chinese Academy of Sciences have introduced peer review, but political influence remains strong in funding: "There is a popular saying in China: 'Small grants, big review; medium grants, small review; big grants, no review.'"[119]

New product definition and innovation require creativity, experimentation, and risk taking. In the words of WebEx CEO Min Zhu, "Silicon Valley is the technology leader and the center for real innovation because it supports the growth of start-ups. New firms cannot grow this fast in China or India. The most powerful model is a truly international company that combines the creative ideas and architectures that are developed in the United States with the ability to quickly implement them where skill is less expensive: both need to be scaleable."[120]

China has built a competitive advantage on a combination of first-world infrastructure and low-cost skill. Its education investments are unprecedented: between 1995 and 2003, enrollment in high school rose from 51 million to 132 million; the annual number of university graduates doubled between 2000 and 2004 to 1.9 million and is expected to double again by 2008 to 3.8 million. This may help to explain the labor shortages in the low-wage factories in south China's Pearl River Delta, where the working conditions and pay for junior-high graduates have not changed in a decade.[121] Even Shanghai now faces high and rising labor costs as multinationals and domestic companies compete for a limited supply of skilled technicians and experienced engineers. Turnover among engineers in Shanghai, for example, is now 20–30 percent per year

The experience of UTStarcom suggests that China is large enough to justify the development of new architectures and products for developing-economy markets, rather than for export to wealthier economies. Products like the *xiaolingtong* use off-the-shelf technologies and address the distinctive needs and resources of lower-income, often rural, markets that are home to two-thirds of the world's population. While the Chinese government is focused on competing with the leading global economies, these underserved markets provide opportunities for Chinese producers to innovate—while potentially also raising living standards for the 4 billion people who are currently living in poverty and minimally integrated into global markets.

7 ❖ IT Enclaves in India

In 1993 Sanjay Anandaram moved to Silicon Valley to manage projects for India's Wipro. The Bangalore-based company, virtually unknown in the United States at the time, counted well-known SV technology firms among its clients—but Anandaram refused to share their names. In his words, "U.S. companies don't want to disclose that their software work is being done in India."[1] A decade later Anandaram was traveling between Silicon Valley and Bangalore as a partner for JumpStartUp, a venture capital firm specializing in early-stage, cross-border technology start-ups in India and the United States.

India's earlier reputation as a technological laggard can be traced to 1978, when IBM angrily withdrew from the country rather than accept local shareholders as required by the Foreign Exchange and Regulation Act. India's domestic IT industry quickly became isolated from its global counterparts by high tariffs, restrictions on computer hardware imports, and the bureaucratic controls of the "license Raj." When the Indian economic environment began to open up following the election of Prime Minister Rajiv Gandhi in 1984 and the subsequent macroeconomic liberalization, U.S. corporations turned to Wipro and its counterparts for programming support to address the Y2K crisis (adapting computer systems to the new century), but most still viewed Indian workers as high-tech coolies.

Only five years later India was accorded the status of an IT superpower by the international media, and the southern city of Bangalore claimed the distinction of being the "Silicon Valley of Asia." This change—the rapid transformation of India's reputation and the growth of domestic confidence—is one of the most visible results of the creation of a technical community linking the United States and India. The high-profile successes of several of Silicon Valley's Indian entrepreneurs and their promotion of regions like Bangalore in the 1980s and 1990s helped to transform the way the world viewed India as well as the way Indians viewed their role in the global economy.

These changes had been underway for several decades, beginning with the emergence of the Indian software industry in the 1970s and the revival of investment by U.S. technology firms like Hewlett-Packard and General Electric in the 1980s. Although U.S.-educated engineers rarely returned to India permanently in these years, many used their positions in large corporations to champion India as a location for a software development center or to promote the use of Indian firms as outsourcing partners. Along with the process of market liberalization that began in 1991, these early successes changed the world's perception of the Indian business environment.

India's software companies grew initially by providing low-end software coding and maintenance services to U.S. corporations at very low cost. In the 1990s, they began to manage outsourced software projects. These export-oriented Indian software services firms, along with a growing concentration of multinational development centers in places like Bangalore, grew at a rate of more than 55 percent a year in the late 1990s, compared to 6.6 percent for the economy as a whole.[2] Yet even Bangalore, where most of the activity was concentrated, lacked the institutional and social environment that would merit comparison with Silicon Valley.

The turn of the century marked a turning point for the Indian technology industry. For the first time a small but significant cohort of U.S.-educated Indian engineers began to return home, and "brain circulation" between the United States and India accelerated. This reflected the newfound confidence (and wealth) of Indian entrepreneurs in regions like Silicon Valley; the post-Y2K recognition in the West that Indian firms offered high-quality services, not just cheap labor; and the increasing openness of Indian policymakers to the advice of the non-resident Indian community. The development of Silicon Valley–style venture capital and other professional services, and the emergence of cross-regional start-ups with headquarters in the United States and development in India, contributed to a deepening of the ecosystem for entrepreneurship in regions like Bangalore.

The rapid growth of the software industry and, more recently, of the outsourcing of such business processes as customer service, technical support, and back-office services has made India the envy of other developing nations. Yet even more than in China, growth remains highly localized, and technology businesses have had limited impact on the Indian economy. Bangalore and its smaller counterparts are fast-growing, wealthy enclaves in a very poor, mostly rural, economy with a national per capita income of less than $1.50 per day.

The software industry's outward orientation (three-quarters of its output is exported) contributes to this enclave effect, as do close ties to Silicon Valley and the backwardness of Indian infrastructure, including expensive and unreliable electric power, roads, rail, airports, and ports. Four decades of state-directed autarkic development strategies have left a residue of tariffs, inflexible labor and bankruptcy laws, and inefficient financial institutions.

Perhaps the most significant barrier to diffusion of the benefits of IT development in India remains its social institutions, particularly the educational system. The IT industry has benefited directly from post-Independence Indian investments in the creation of a small tier of world-class institutions of higher education: the Indian Institutes of Technology and Management as well as the Indian Institute of Science. The rest of the educational system, from elementary school through high school, remains so thoroughly inadequate that the Indian adult literacy rate of 58 percent (46.4 percent for women) in 2001 placed the country below the average of the U.N.'s "low-income" countries. This stratification of the educational system contrasts to the relative equality of educational opportunities in China, where adult literacy was 85.8 percent, and slightly higher for women than men. India's limited health care system results in an infant mortality rate of 67 deaths per 1,000 live births (compared to China's 31), and life expectancy at birth of 64 years (versus 71 years in China).[3]

THE ORIGINS OF INDIA'S HIGH-TECH INDUSTRY

Tata Consultancy Services (TCS), a part of one of India's oldest and largest conglomerate business houses, the Tata group, pioneered the Indian software development and consulting service businesses in the late 1960s. The managing director of the Mumbai-based firm, F. C. Kohli, foreshadowed the potential of IT for India in a 1975 speech:

> Many years ago, there was an industrial revolution. We missed it due to [factors over] which we had no control. Today there is a new revolution—a revolution in information technology, which requires neither mechanical bias nor mechanical temperament. Primarily it requires the capability to think clearly. This we have in abundance. We have an opportunity even to assume leadership. If we miss this opportunity, those who will follow us will not forgive us for our tardiness and negligence.[4]

Software

As in Taiwan and China, the opportunities and relationships created by access to university-level engineering education in the United States were critical to the origins of the Indian information technology industry. Kohli, an electrical engineer with graduate degrees from MIT and the University of Waterloo, generated TCS's first outsourcing contract on a trip to New York in the early 1970s for a board meeting of the Institute of Electrical and Electronic Engineers. Several years later S. Ramadorai, a University of California graduate and now CEO of TCS, established a New York office to help grow the American customer base. TCS grew and gained a reputation for professionalism and quality in the 1980s by providing low-end computing services—mainly software coding and maintenance of mainframes—for large corporations like Burroughs, American Express, and Sega. By the end of the 1980s TCS was the largest Indian software and IT services company, a position it retains today.

An American embargo on the export of high-technology equipment to India (long considered an ally of the Soviet Union), combined with severe import restrictions on hardware, meant that early outsourcing work was done by Indian workers at client facilities in the United States, or, as the somewhat confusing Indian terminology puts it, on-site. (The Indian industry is so oriented to other countries that services performed in a location remote from the client's premises, even at home in India, are called offshore.)

The pejorative term for providing software services on-site at client facilities, "body shopping," reflects the reality that this work amounted to little more than a highly lucrative labor cost arbitrage at a time when Indian workers were willing to work in the United States temporarily and to be paid in rupees at 5 to 10 percent of the wages of their U.S. counterparts. The barriers to entry were quite low: all that was needed to become a body shopper was a Rolodex file of the names of programmers in India. However, the developmental potential for India was minimal because this system generated only scattered projects and exacerbated the brain drain, since a large proportion of Indians sent to the United States temporarily to work on contracts managed to find jobs that allowed them to stay.[5]

When Prime Minister Rajiv Gandhi was elected in 1984, he initiated policy reforms that reversed the regime of technological self-reliance and encouraged foreign investment in India. The Computer Policy of 1984 recognized software as an "industry" entitled to investment and incentives as well as reduced import tariffs on software and PCs (from 100 percent to 60 percent). In 1986 the Computer Software, Development, and Training Policy liberalized access to technology and software tools, invited foreign investment, and supported access to venture capital. While this removed the main barriers to the growth of technology industries, it did not represent positive support of the type found in Taiwan at this time. Indian software companies remained isolated from the global economy because of bureaucratic import procedures, high duties, and limited access to foreign exchange. In addition, the absence of reliable telecommunications links ensured the continuation of on-site services for many years.

Less visible aspects of Rajiv Gandhi's tenure were at least as important as these policy changes. In the early 1980s government officials invited Radha Basu, a U.S.-educated South Indian manager from Hewlett-Packard who was advising the Indian Department of Electronics, to invest in India. In 1985 she set up HP-India, which was one of the first two foreign software development centers in Bangalore. Gandhi's technology advisers also personally recruited General Electric (GE) chairman Jack Welch to sell software and to invest in India. GE quickly became the largest foreign investor in India's software and technology industries, providing both credibility and business skill to the fledgling industry. In the 1990s GE was the largest client for both TCS and Wipro and, in a manner reminiscent of Taiwan's OEM relationships, helped its vendors develop a focus on business efficiency and cost-cutting.[6]

Hardware

An indigenous computer hardware manufacturing sector, stunted by years of protection and limited access to foreign technology, began to develop in India at the same time as the software industry. When Azim Premji, an engineering student at Stanford University, was forced in 1979 to return home prematurely to run the family-owned cooking oil business, he recognized the opportunity created by IBM's recent departure. Premji changed the name of his company from Western Vegetable Oil Company to Wipro and began to tap India's underutilized engineering talent to develop and manufacture minicomputers.

Wipro quickly established itself as India's leading computer maker and was able to attract the country's top engineering graduates because it offered an opportunity to do original research and development. By 1985 Wipro had developed its own Intel-based PC and established distribution of its products in Asia through a partnership with Taiwan's Multitech (later Acer). The company gained attention as a global leader in microprocessor design and development in 1986 after introducing a prototype 80386-based system only two months after Compaq and ahead of IBM. In 1989 Intel selected Wipro as a "beta test site" for systems based on its 80486 microprocessor.

The opening of the Indian market that began in 1991, however, soon forced Wipro and its local competitors to abandon computer development and manufacturing. Indian producers, crippled by the small size of the domestic market and lacking significant foreign customers, were unable to compete with Taiwan's increasingly efficient and flexible manufacturers. Wipro maintained a small semiconductor design division in Bangalore; most other Indian companies became direct or indirect dealerships for foreign-brand computers and related products. Indian IT hardware revenues stagnated as a result, and the industry to this day has negligible exports.

Wipro's executives, watching both TCS and the more recently formed Infosys (founded in 1981 in Bangalore), recognized that there were lucrative opportunities available in software services outsourcing. They established Wipro Technologies in 1989 as the software arm of the business and adopted the on-site model—sending teams of Indian engineers to the United States to perform low-value-added services such as legacy systems support and software coding, testing, and maintenance.

A decade later Vivek Paul, a U.S.-educated executive from General Electric (GE), was recruited to run Wipro Technologies. One of Paul's first initiatives in 1999 was to move the division's headquarters to Silicon Valley to establish a presence in the market. He began to commute regularly between Mountain View, California, and Bangalore, India, and reports that the day-to-day interactions with customers allowed Wipro employees to identify new market opportunities quickly and convey them rapidly to India. Paul also transferred key elements of GE culture to Wipro, including a focus on speed and execution, engaging customers, empowering workers, and eliminating hierarchy. Under Paul's leadership, Wipro Technologies grew from a $150 million company in 1999 to $1.4 billion in 2004. The business now generates two-thirds of Wipro's revenues and more than 90 percent of its profits.[7]

THE MULTINATIONAL CONNECTION

Soon after Indian businesses like TCS began exporting software services to American customers, U.S. corporations pioneered the first software development centers in India. In 1985 Hewlett-Packard (HP) and Texas Instruments (TI) established development centers in Bangalore at the invitation of government officials. Radha Basu, who set up HP-India, acknowledges that the bureaucratic restrictions and inadequate infrastructure of her home country at the time were "anything but inviting." When Indian officials claimed to have "a single window of clearance," she would joke: "Where's the window?" Basu, who operated out of her dining room with a telex machine for most of her first year, reports that in addition to many anticipated difficulties such as getting a phone, energy, and a water pump, she was unable to convince customs agents that it was possible to export software electronically; for five years she was forced to copy data (which had already been transferred electronically) onto tapes that she shipped to a warehouse in the United States in order to ensure that HP-India received tax breaks for exporting software.[8]

HP-India was an unambiguous success story. However, Radha Basu accepted substantial professional risk and personal hardship in an undertaking that was considered foolhardy by some. She reports being motivated by her strong identification with and personal commitment to India's development, and by the fact that HP would not have established the center without her efforts. Basu used her credibility with both HP management and Indian policymakers to gain the trust and confidence of both sides, and she played a critical role in laying a foundation for cooperation between Indian government officials and a subsequent generation of American multinationals.[9]

Doing business in India in the 1980s was frustrating even for a native-born manager. Basu recalls meeting repeatedly with officials from several different ministries (including Electronics, Industry, Telecommunications, and State) to convince them of the need for reliable, dedicated telecommunications capabilities, since private satellite facilities were prohibited at the time. These bureaucrats were often hostile to foreign business, resentful of "non-resident Indians" (NRIs), lacked understanding of technology, and rarely communicated with one another. The problem was only resolved in 1986 when TI donated the satellite communications technology and equipment to the Indian government, which in turn allowed U.S. firms to use the excess capacity. Even then, the process involved removing or breaking twenty-five different regulations.[10] In 1989 the public telecommunications company, VSNL, commissioned a direct 64-kbps satellite link to the United States for software exporters, but most foreign firms eventually established private satellite connections.

The successes of HP and TI marked a turning point for the relationship between American technology businesses and India. News spread rapidly among expatriate Indian managers, who recognized the potential of India's large, low-cost, English-speaking skill base and began to consider India as a possible location for their corporations' software operations. Khailash Joshi spearheaded IBM's reentry into India in 1991, and achieved government approval for the first joint venture between an Indian technology business (TCS) and an American multinational. Joshi was general manager for IBM in Lexington, Kentucky—now Lexmark—at a time when the company was, in his words, "completely white and blue-blooded." Like Basu, he acknowledges that his commitment to India motivated him to risk his professional reputation and invest time and energy into the background research needed to make a convincing case to IBM's board.[11] It is difficult to imagine a non-Indian manager investing as much in the effort, particularly at that time.

The Power of Example

Radha Basu and Khailash Joshi became well known among the Indian technical community in the United States for bringing their companies' operations to India. Both returned to Silicon Valley and started their own companies, and both have received awards and other recognition for their contributions. The level of their public support and personal investments in promoting Indian technology development is unusual. However, many other expatriate Indians also actively pursued corporate initiatives in India, combining a personal commitment to India with bottom-line efforts to improve business efficiency.

In 1992 Ranga Puranik established Motorola's India Software Center after researching the TI experience, promoting a location in India to senior management, and agreeing to relocate to India along with two other Indian managers to run the center. The growth of the

India Center, which started with only ten engineers, depended entirely on their ability to attract projects from the (often wary) managers of Motorola's autonomous units. Many of these managers had understandable concerns about sending business to a distant country with primitive infrastructure, an incomprehensible bureaucracy, and corrupt, unresponsive officials. They worried that their engineers might not be able to communicate effectively with their counterparts in India, and they faced understandable fears among their domestic colleagues about losing jobs to India. Not surprisingly, the India Center's first five or six contracts were sponsored by Indian-born managers, either directly or through recommendations to their superiors.[12]

There were also failures. In 1996 Ashok Khosla, a second-generation Indian-American, enthusiastically accepted responsibility for setting up an Apple Development Center in Bangalore. After two extremely difficult years he recommended that Apple shut down the facility, which it did, and he advised other local companies not to open branches in India but rather to recruit from there. When asked about the problems, he responded:

> Where do you start? Water, power, roads, telephones. . . . Power is really difficult to get in Bangalore. . . . When you first get there your entire life mission is to find a source of power. . . . Then you have to get telephone access . . . getting telephones is really a challenge . . . you have to bribe. . . . The average for a resident who pays a bribe is six months to a year . . . for someone who doesn't pay the bribe it could be 10 to 12 years. If you want anything done you have to pay a bribe.[13]

He went on to describe other difficulties: it took four people negotiating for three months to get a satellite antenna put up on a tall building, and he had to pay a $30,000 installation fee and $10,000 a month for a 64-kbps connection. The advantages of a native-born

Indian manager seem clear from Khosla's experience, although both TI-India and DEC succeeded with American managers.

Talent, Talent, Talent

Silicon Valley companies continued to set up development centers in Bangalore during the 1990s as they began to recognize that India offered a source of scarce technical talent as well as cost savings. By the late 1990s HP-India had responsibility for a wide range of software development, including operating systems, networking, object-based applications, and security software. Radha Basu emphasizes that senior management at HP no longer saw the India development center as a source of low-wage labor: "The combination of high talent with effective costs is what attracted us to India. I have to tell you now that it is talent, talent, talent."[14]

In the late 1990s Oracle-India developed a new operating system for an Oracle network computer. Nimish Mehta, who helped establish the center in Bangalore in 1991, acknowledged that his motivation was "absolutely based on the notion of giving back." He explains:

> [We] started off with cost reduction, you know, the "cheaper there" argument. But you look at that a little bit and think this makes no sense because if the only objective is cost-reduction, then you ought to have all your programmers in one location. . . . The cost of bad communication, missed deadlines because you meant one thing when you said another thing, is far greater than the cost of $40–50,000 a year for a programmer. So if the cost is the issue, have them here. If talent is the issue, then you need to have some people there [India] as well.[15]

Like many of its counterparts, Oracle established its own private satellite link to Bangalore and began using new software tools to accelerate information flows and facilitate a 24-hour programming cycle: "We've put some tools here that allow us to specify what to build at a higher semantic level. . . . You can only do it if there's not a lot of context that has to be carried between the people that are working there and the people that are working here. The context now has to be contained in the tool and not in the person's brain." The introduction of increasingly sophisticated software tools to support long-distance collaboration in the late 1990s allowed companies like Oracle to perform a growing proportion of their software development work overseas.

Support from Below

The thousands of Indian engineers working in American technology companies played less visible, but often critical, roles in ensuring the success of operations in their home county. Atul Vijaykar, who advocated strongly for Intel to commit resources to market and engineering development in India, observed: "What general management wants is . . . not only [to reduce] cost but also, above all, they want the project to succeed. [Within Intel] there are many Indian senior managers in various engineering positions who have, in their own indirect ways, supported projects being done offshore in India."[16]

Indian-born engineers took the initial risks of sending complex projects to India, and frequently took responsibility for ensuring their final quality as well. They coordinated the remote software development projects, adapted their schedules to accommodate the 12-hour time difference, worked overtime to help meet pending deadlines, and in some cases even relocated to India temporarily to oversee or troubleshoot these projects. When Cisco began a partnership with the Indian firm HCL, a team of fifty Indian engineers volunteered to relocate for the early stages of the collaboration. Thousands of other Indians, many of whom had worked in the United

States for decades, have moved to India temporarily since 1990, either to establish or manage corporate operations or to manage partnership.

Indian managers also actively promoted the upgrading of technological capabilities in their home country. In 2002, Intel sent an entire flagship, next-generation microprocessor development project to India. Vijaykar says confidently, "It would not have happened without a senior [Indian] manager within the business unit willing to move to India and lead the overseas operations." He is proud of Intel's role in transferring technical know-how to India: "You'll have over the next five to ten years people trained in circuit design and hardware design where a few years back, there were none."[17] Intel's Bangalore campus is now leading worldwide research on 32-bit microprocessors for servers and wireless chips. Engineers at Texas Instruments' chip design operation in Bangalore have earned 225 U.S. patents and developed a digital signal processing chip that was commercialized worldwide. Recognition of the growing sophistication of the research and engineering work done in India for well-known corporations has not only increased the confidence of the Indian community but has also provided domestic Indian software companies with an alternative model to low-end on-site outsourcing.

Bangalore is now home to more than 100 multinational development centers, and it is unlikely that this cluster would have developed so rapidly without the contributions of Indian-born engineers from Silicon Valley and elsewhere. Kavita Goswamy concludes her study of IT development in India unambiguously: "Expatriates helped India create its niche in the global IT market . . . in all levels of management they not only supported doing business in India, they have taken the initiative to champion it. . . . The motivation derives from a personal and patriotic attachment to India, despite living in the U.S. for many years, so that Indians go beyond the normal requirements of their jobs to make the India center successful."[18]

COMPETING FOR SKILL

The brain drain remained a harsh reality for India for decades. Even in the late 1990s, when many overseas Chinese were experimenting with returning home to start Internet businesses, Indian engineers sought out ways to get to the United States, and once there, they almost never chose to return to India permanently. By 2002, however, the combination of recession in Silicon Valley and growing professional opportunities in India triggered the first sustained interest in returning home on the part of U.S.-educated Indians.

The Indian software organization NASSCOM reports that some 5,000 overseas Indian technology professionals with more than five years of work experience moved back to India from the United States in 2002–2003. It is hard to verify these numbers, but one indication is that a 2003 job fair in the Bay Area sponsored by *siliconindia* attracted 1,200 Indians who were considering returning home. Most potential returnees were looking for jobs in established multinationals like HP or Texas Instruments. Intel reported in 2003 that nearly 10 percent of the 1,400 engineers in its Bangalore center were repatriated Indians who had spent significant time abroad. Even expatriates with U.S. graduate degrees who had little or no job experience returned to India in greater numbers, pushed by a tough job market and the high cost of housing in places like Silicon Valley, and pulled by growing demand for their skills in places like Bangalore and Mumbai.

The sentiments of these returning Indians echo those of Chinese technology workers who returned home—the sense that there are more opportunities to grow professionally as well as personally in India than in the United States, the feeling that India will provide higher-quality work opportunities and more career growth, and the ability to build a better lifestyle in India because of the lower cost of living and the extended family support systems. In the words of

Vinod Gopineth, who returned to India to work for a semiconductor start-up after a career in the United States that included a position at IBM, "When you are looking to ride a wave, there can be none better than the one at home. Your options are better. There is more pride. The challenge is greater. You spend all your time in the U.S. just proving yourself. Now here was an opportunity to prove, not just yourself but your country to the rest of the world."[19]

Upgrading Capabilities

The growing presence of multinational development centers transformed competition in the Indian software industry and forced domestic firms, for the first time, to compete for skill. As Infosys CEO N. R. Narayana Murthy explained, "The multinationals were a godsend for Bangalore. They created tremendous competition—not for customers but for employees. Competition forced us to scale up our operations and build a local infrastructure so that we could attract and retain qualified employees. . . . Our challenge was to attract, train, empower, and retain the best and brightest professionals."[20]

The multinationals not only created competition for talent, they also demonstrated to Indian companies the possibility of offshore service provision (that is, working directly from facilities in India). The removal in the early 1990s of import licenses on computer equipment and industrial imports, and the establishment of Software Technology Parks (STPs), made it feasible to provide services from India. The Department of Electronics' STP program provided export-oriented software firms in designated zones with access to high-speed (64 kbps) telecommunications links and reliable electricity, as well as tax exemptions. Once Indian firms like Infosys and Wipro began establishing their own facilities, the proportion of software services provided offshore grew quickly. The share of software output originating in India grew to 59 percent in 2004, from only 5 percent in 1990. Moving operations back home allowed Indian software firms to invest for the first time in building workforce skills and in accumulating the organizational and process knowledge that creates barriers to entry. It also allowed them to take orders for complete programs, to work directly for final clients, to bill by the project rather than by labor hours, and to design and build their own workspaces.

Training was central to these capability-building efforts. Wipro began providing management, technical, and consulting training internally in the late 1990s. The firm established Wipro University, which provided a six-month training course for recruits from nonelite universities (a more and more common practice as competition for skill intensified). Wipro also established an Academy of Software Excellence to provide a three-year orientation to engineering for math and science graduates, and worked with local universities to build appropriate curricula. In 2001 Infosys invested $7.3 million in a training institute; by 2004 the firm claimed to run the biggest corporate training facility in the world, with 4,000 students per session and three sessions a year.[21]

The leading Indian firms invested heavily in facilities as well. Infosys used the cash from its successful IPO in 1993 to build a 70-acre university-like campus in Bangalore. The facility is open 24 hours a day, allowing flexible work hours; it provides technologically advanced equipment and an attractive work environment; the restaurant provides free meals and coffee for employees; and there is a gymnasium with basketball and volleyball courts, a swimming pool, and a sauna. According to founder Narayana Murthy, the goal was to create an environment that was familiar and comfortable for foreign (mostly American) customers, as well as to increase loyalty among the workforce at a time when annual employee turnover in the city was higher than 20 percent. Murthy also modeled governance mechanisms after American practice: "Our view was that we needed to learn from successful experiments like Silicon Valley . . . for example, we implemented India's first stock option plan for employees and we created a very open and egalitarian environment . . . We also constantly benchmark ourselves against the best in the world."[22] The firm abandoned the traditional Indian hierarchies, treating both customers and employees like partners, and focused on providing service and innovating.

Once they had their own premises, the large Indian firms increased productivity by improving project management skills, developing more formal and effective process and quality methodologies, and building organizational capabilities. These companies became increasingly efficient at managing large-scale customized application development and maintenance while maintaining quality and ensuring the privacy and security of their clients' work. There was still, however, the problem of reputation. In the words of Infosys cofounder Murthy, "In the early '90s, when we went to the United States to sell our services, most chief information officers didn't believe that an Indian company could build the large applications that they needed. The CIOs were very nice to us, of course. They offered us coffee or tea, listened to what we had to say, and then said, 'Look, don't call us—we'll call you.'"[23]

The Y2K crisis, combined with the skill shortage in the United States due to the Internet boom of the late 1990s, drew attention to the skills of Indian software engineers. The Indian IT magazine *Dataquest* reported that "Y2K—the biggest bogey in software history— . . . triggered off a chain of events. Exports grew on the strength of Y2K and never looked back. The local training industry boomed partially because of it. . . . Locally, full-page government ads in leading dailies asking companies how to become 'Y2K compliant' helped the paranoia along. By the end of 1999 the industry was on an all-time high. IPOs of software companies were getting oversubscribed several times."[24]

The successful Indian completion of the Y2K projects for leading American financial and retail services like Citibank enhanced confidence among large foreign customers and produced a substantial increase in customers and orders. By specializing in services for banking, financial services, and insurance (which accounted for 40 percent of industry revenues in 2003), Indian companies accumulated valuable domain knowledge as well as a reputation for quality. These process and quality improvements are reflected in the superior performance of Indian firms on the Carnegie-Mellon Software Engineering Institute Capability Maturity Model (SEI CMM), which promotes best practices in software development.

In 1993, Motorola's India Center became the first software operation in the world to achieve a CMM Level 5 quality rating, and Motorola's CEO traveled to India to award a gold medal to the one hundred employees who received the Level 5 certification. The certification established the reputation of Motorola India for quality work, which in turn attracted a growing volume of assignments from non-Indian managers in the United States. It also created a powerful demonstration effect for domestic Indian companies, who themselves began pursuing CMM certification.

By 1999 ten Indian firms, including Wipro and Infosys, had achieved CMM Level 5 certification, compared to only six companies in the United States. In 2003 there were sixty Indian companies with Level 5 certification—accounting for three-quarters of the world's total.[25] By contrast, only six Chinese software companies were certified at Levels 4 or 5.[26] This rating allowed Indian firms to continue increasing exports for more than a decade in the face of labor shortages, rapidly rising wages, and a severe recession in their leading market. By 2005, more than half of the Fortune 500 companies were clients of the Indian software industry.

The Bangalore Boom

The early location of software businesses in India has become self-sustaining. Bangalore remains the dominant hub for the IT services and software industry, accounting for about one-third of software exports and employment since data first became available in the mid-1990s. Infosys was founded in Bangalore in 1981 and was quickly followed by other leading Indian and U.S. companies, including Wipro, Texas Instruments, Hewlett-Packard, and Digital Equipment Corporation. Like Silicon Valley three decades earlier, Bangalore continues to attract new IT-related ventures (and thus employees) in spite of fast-rising wages, inadequate infrastructure, and worsening traffic congestion and pollution. By 2004 the city was home to some 1,200 IT-related companies and over 260,000 employees. Industrial growth continues to drive accelerated population growth. In 1981 the population of Bangalore was around 2 million; by 2001 it had more than doubled to 5 million; and it is forecast to reach 8 million by 2010.[27]

As human resource constraints increase in mature locations like Bangalore, where attrition rates are 20–30 percent per year (and up to 60 percent for firms doing BPO) and wages are rising by almost 15 percent per year, India's IT services and software firms are seeking alternative, lower-cost locations. The Delhi and Mumbai areas face similar labor supply constraints, leading firms to establish new operations in other less costly but talent-rich cities, including Pune, Chennai, Hyderabad, and Kolkata.[28]

This concentration of the software industry in the southern peninsula has deep historical roots, paralleling the concentration of India's public engineering resources (both R&D establishments and publicly funded engineering colleges). In 1911, the Maharaja of Mysore proposed state involvement in the promotion of engineering and higher learning at the Mysore Economic Conference. India's first Prime Minister, Jawaharlal Nehru (1947–1964), later dubbed Bangalore India's "City of the Future" and sought to turn it into the nation's intellectual capital by locating government research laboratories and public-sector manufacturing establishments in the area.[29] Bangalore's state-owned technology businesses include Indian Telephone Industries, Hindustan Aeronautics Limited, and Bharat Electronics Limited; and its scientific research capabilities include the prestigious Indian Institute of Sciences, the Indian Space Research Organization, and the National Defence Research Laboratories.

Karnataka (where Bangalore is located) and nearby Andhra Pradesh (Hyderabad) were also the first states in India to allow the establishment of private engineering colleges. There are currently 77 approved engineering colleges in Bangalore that produce 29,000 graduates annually; and the four southernmost provinces of India together are home to more than 400 engineering colleges, accounting for two-thirds of India's total annual engineering student admissions.[30] Karnataka also set uniform educational standards and thus ensured a minimum threshold of quality. Ashok Desai argues that this was one of the major factors behind Bangalore's rise as an IT center. He also points out that Bangalore was open to outsiders, as were Mumbai, Poona, and Delhi, while other parts of the country such as Calcutta that remained more provincial failed to develop as quickly.[31]

One of the few arenas that has drawn the otherwise competitive and autarkic Bangalore companies together is process and quality improvement. In the 1990s the largest local firms, under the leadership of Infosys, created a Quality Association that brought together quality managers to share best practices for measuring performance, reducing defects and cycle times, cutting costs, and improving quality. This collaboration helps account for the high level of commitment in India to the SEI CMM model—and ultimately for their higher productivity levels.

Adapting Education

India's educational system, both public and private, has expanded quickly in response to the needs of the growing software and IT industries. While the six prestigious Indian Institutes of Technology still produce only about 1,800 graduates per year (and about 30–40 percent of their engineering graduates settled abroad during the 1980s and 1990s), the country now has more than 250 universities and engineering colleges that provide the computer-related degrees and diplomas most sought after by the software industry. These programs responded, albeit belatedly, to growing demand for skill by increasing their output considerably. In 2001–2002, there were about 130,000 new IT graduates from universities and engineering colleges in India, more than double the number just five years earlier.[32] The government also founded two new undergraduate institutions—the International Institutes of Information Technology in Bangalore and Hyderabad—in the late 1990s, and foreign investors (many from Silicon Valley) established a new business school in Hyderabad. India still produces very few Ph.D.s, however, which will continue to limit the development of domestic research capabilities.

Thousands of private training institutes also emerged across India in the 1990s to provide training of varying levels of quality and rigor. Two major national training providers, NIIT and APTECH, offered software training through franchises and awarded their own degrees. They also served as a base for training and certifying engineers in foreign software from Microsoft, Oracle, Cisco, and IBM. Many other "fly-by-night" operators emerged to take advantage of the demand for computer training. Although the quality of these private institutions is often debated, they have trained hundreds of thousands of Indians in basic computer literacy and programming skills at a time when public institutions were slow in responding to popular interest. While training has increased in recent years, demand has skyrocketed as well. Labor turnover in regions like Bangalore reached 30 percent during the 1990s as workers sought to maximize their earnings in a very tight market; this contributed to wage increases on the order of 25 to 40 percent annually.[33]

IT-Enabled Services

The growth of the IT-enabled services (ITeS) industry in India reflects the outsourcing of another specialized segment of the IT services value chain. The sector, also referred to as Business Process Outsourcing, or BPO, was pioneered by multinational corporations like British Airways, General Electric, and Citigroup in the late 1990s. These firms succeeded in cutting costs substantially by investing in offshore centers for low-skill and labor-intensive, but increasingly automated, processes such as customer service, call centers, and accounting, billing, and payment services.

The success of these early efforts, based on the large wage differential between India and the West, triggered a wave of foreign investments in India as well as the formation of a new generation of indigenous Indian businesses. Domestic start-ups such as Spectramind, Evalueserve, and 24/7 were often founded by former managers of captive multinational centers with the aim of providing specialized third-party BPO services. The substantial cost savings in these operations in turn created competitive pressure in many industries to shift routine, back-office operations to lower-cost locations like India. A McKinsey Global Institute report on ITeS estimates that for every dollar that is moved offshore, the company gains 58 cents in net cost reduction for a comparable or superior level of service.[34]

Expatriate Indians once again played a critical role in seeding this new industry. Raman Roy, who had helped establish American Express in India earlier in the decade, was hired by GE in 1996 to begin outsourcing its financial services. He faced tremendous resistance from business unit leaders when he sought to set up a call center in India to process real-time transactions, not simply batch transactions. When told by a senior executive that call centers would never happen out of India and that the company would not finance individual fantasies, Roy says: "It was nationalistic pride that drove me to see that a call center came into being. I had no budget to experiment. But four members of my management team and I were fired with a kind of missionary zeal to prove what we could do in India." After convincing the equipment manufacturers to loan them the setup for a five-hundred-seat operation and identifying a potential customer willing to fund part of the needs for the proof of concept, Roy says that "we demonstrated that we could match, if not better, the voice work happening overseas. It was a runaway success. Today, I believe that more than 50 percent of GE Capital's India revenue comes from its call centers."[35]

Roy also used GE's "six sigma" approach to quality in order to show what could be achieved in India, and he claims that the unit produced better quality on assignments than they believed was possible: "Errors per hundred and per thousand, turnaround times, accuracy, understanding of what the customer wanted, and delivering that, promising to pay versus actual payment—we measured the quality of our output on all these dimensions, and of course the parameters differed from process to process. I found that, in general, we delivered superior quality and bettered existing benchmarks on many of these parameters."

Nevertheless, India remained a challenging place to do business. In 2004 Raman Roy observed:

> There are certain aspects of India that the local [Indian] managers learn to live with. We grow up with the fact that our phones will not work for part of the time. We grow up with the fact that after you order equipment, there can be a six-month delay before it is delivered. We learn to function in spite of the fact that processing of approvals will not be at lightning speed. We cut our teeth on battling the poor infrastructure and related obstacles and learn to work through and around these problems . . . handling the Indian

> work force, which has its own idiosyncrasies . . . The pain has decreased over the last eight years . . . but it has not disappeared. . . . [Some] areas have unbelievable levels of bureaucracy. For instance, we just set up a center in Pune, a city in western India. I counted the number of signatures required on various documents. There were I think 119 places that I signed to get approval to start an operation there.

The Indian ITeS business is still young. It has grown very quickly, however, and could have a much broader impact on Indian development than software services because of its wider range of skill requirements, the advantage of English language skills, and the potential for efficiency improvements. In 2002 BPO attracted 15 percent of total foreign direct investment in India, including investment by leading IT service firms like Accenture and EDS, and it accounted for 10 percent of all exports. Improvements in the telecommunications infrastructure as well as official incentives (including a waiver of the 35 percent tax on corporate profits and exemption from many of the country's labor laws regarding hours and overtime) helped attract these investments.[36] The cash-rich large Indian software service companies also expanded their offerings to include BPO services, hoping to leverage existing customer relationships. Wipro, for example, acquired Raman Roy's start-up, Spectramind. Perhaps as a result, the industry has largely followed the locational pattern of the software sources industry, and is concentrated primarily in Bangalore, Delhi, Pune, Hyderabad, and Mumbai.

The early BPO activity in India was primarily "bums-on-seats"—that is, intensive back-office services such as call centers staffed by Indians with cultivated American or European accents. This is changing quickly, however. The leading BPO businesses in India are adding more value, for example by using IT for data analysis, data mining, and data management. Human resources (HR) outsourcing services began providing faster, better, and cheaper payroll and personnel record maintenance, and over time this has expanded to include remote personnel management, resource management, and talent management. Likewise, financial and accounting outsourcing have shifted from travel claims processing and bill payment applications to accounts payable and accounts receivable systems that incorporate vendor and customer master management, voice and e-mail support to collections and payments queries, and so forth.

Indian vendors now actively promote strategic services such as systems integration, network consulting and management, market research, engineering and design, and translation and transcription; and they talk of future opportunities to provide higher-value-added services such as logistics management, consulting, legal and medical advice, and distributed product development. Pipal Research, a small Indian firm set up by former McKinsey employees, provides research services for consultants, investment bankers, and company strategy departments. Wipro Spectramind is moving from call centers to providing accounting, insurance, procurement, and product liability services.

Emerging Giants

The Indian IT industry grew from $5 billion in 1997 and 1.2 percent of the nation's GDP to $21 billion and 3.5 percent of GDP in 2004. The dominant sector in 2004, IT Services and Software, accounted for more than half of total revenues ($12.8 b.), followed by IT Hardware ($4.8 b.) and ITES-BPO ($3.9 b.). The fastest-growing sector, ITES-BPO, is almost certain to surpass IT Hardware in revenues in 2006. The industry remains heavily export-oriented, with almost 62 percent of total sales for export in 2004, and the dependence on exports appears to be growing. Since 2000, IT exports have grown by over 20 percent per year, compared to approximately

5 percent growth rates for the economy as a whole. The export earnings of the IT sector have contributed to India's accumulation of a balance of payments surplus and sizable foreign exchange reserves.[37] However, the overwhelming export orientation of these producers may prove to be detrimental if local providers fail to serve the domestic market, or even cede it to foreign competitors.

The structure of the Indian IT industry has changed as well, reflecting the accumulation of capabilities by the largest firms with the resources to build their own facilities. While low barriers to entry had once attracted hundreds of small- and medium-sized firms into the software industry, by 2003 five firms—TCS, Infosys, Wipro, Satyam Computer Services, and HCL Technologies—accounted for close to a third of total Indian software and services exports, and the top ten accounted for 55 percent. Subsidiaries of overseas companies accounted for another quarter of exports.

Three giant Indian software service firms—TCS, Wipro, and Infosys—have remained in leadership positions for over a decade. By 2004 each had more than $1 billion in revenues, over 35,000 employees, and, in spite of their large base size, sustained growth rates of over 30 percent annually.[38] These three firms also rank in the top ten global IT services companies in terms of stock market capitalization (all are listed on the New York Stock Exchange), gross profits, and employees. They are also actors in the BPO sector, and provide increasing competition for global consulting and services giants like IBM, Accenture, and EDS. They have all sought to expand their global presence and outlook, in part through acquisitions of foreign companies, as well as by setting up new centers in locations such as China and Eastern Europe.

Indian executives argue that their competitive advantage lies in their ability to effectively leverage the technically skilled, low-cost talent in India and other emerging economies. The Infosys Global Delivery Model divides development tasks into logical components and distributes the components geographically, performing them where it creates the maximum value. Some tasks, such as user training and defining business requirements, must be performed close to the customer, while others are best performed remotely in "scalable, process-driven, technology-enabled development centers located in cost-competitive countries like India."[39] The next generation of strategic services outsourcing, according to Infosys, Modular Global Sourcing, applies the concepts of modularization to business process and IT application and infrastructure services sourcing decision making, implementation, and ongoing management.

The critical advantage of Indian firms relative to the major American IT services suppliers is the distribution of their workforce. While two-thirds or more of the workforce of companies like Wipro, Infosys, and TCS is in India, a comparable percentage of the workforce of western companies like EDS, IBM, and Accenture is located in the United States—even if they have a large presence in India—and this legacy workforce is largely immobile. The different cost structures alone go a long way toward explaining the substantially higher margins of the top six Indian firms compared to the top six traditional outsourcing suppliers. The Indian firms maintained an average net profit margin of 21.7 percent in 2004, in contrast to the average net profit margin of only 4.3 percent for their traditional competitors.[40]

The Indian companies have reinvested their earnings in developing sophisticated tools, methodologies, and processes to improve the efficiency of a globally distributed workforce. They have adopted well-defined hiring and training systems to help them attract and retain the top talent in India; they train employees aggressively and create incentives for excellence (including awarding 25 to 50 percent of an employee's salary based on the performance of the individual, the team, and the company); and they place great emphasis on strong quality and project management processes, along with measurement and benchmarking of a wide range of parameters throughout the development process. Wipro has explicitly modeled its business services processes after the Toyota manufacturing principles, seeking to improve productivity, quality, and efficiency through improvements in work organization, task partitioning, the adoption of software tools, quality programs, and so forth.[41] While foreign competitors can set up development centers, it is unlikely that they will be able to invest sufficiently to develop a workforce with comparable scale, experience, and commitment.

VENTURE CAPITAL AND ENTREPRENEURSHIP IN INDIA

Until recently, India has not been a hospitable environment for technology entrepreneurship. Veteran SV venture capitalist William H. "Bill" Draper III took U.S.-style early-stage venture investing to India in 1995. Soon after establishing a $55 million fund, Draper International, with a colleague in San Francisco and local partners in Bangalore and Mumbai, Draper began to recognize the obstacles involved in working with entrepreneurs in India. The concept of venture capital was new in India, where closely held family businesses dominated both small firms and the large industrial groups like Birla and Tata. It was very difficult to coordinate activities between India and Silicon Valley. Moreover, start-ups with far more limited resources faced the same frustrations as multinational investors: the frequently corrupt and restrictive bureaucracy, the lack of Western-style stock option plans to retain key personnel, the limits of basic infrastructure such as power and water, and the very high cost of connectivity. In the words of Draper's partner, Robin Richards, in 1997,

> From a venture capitalist's perspective, it is 5–10 times slower to start up a company in India than in the U.S., mostly due to the dif-

> ficulty of getting the correct infrastructure in place for production. And since most of the technology market is focused in the West, it is difficult for an India-based entrepreneur to form the necessary business partnerships and find the right management, marketing, and sales talent to join his or her company.[42]

Within two years, Draper and Richards had reoriented the strategy for their fund to focus on investments in U.S.-based firms with a large proportion of their activities in India. When the fund closed in 2000, only one of the India-based companies that had received funding was successful (Rediff Communications), whereas a majority of the U.S.-based firms in the portfolio had achieved liquidity. The timing was fortuitous, to be sure, but Draper attributes the successes of the fund to the combination of technology development in India and business, marketing, and sales operations "near the action" in the United States. Some of the firms in the fund were in the United States to begin with, but others were in India, and Draper brought them to Silicon Valley. According to Robin Richards, they learned about some of their best U.S.-based deals in India: "You could be sitting in Bangalore and hear about 10 great deals in Santa Clara that you could be a VC investor in, but you'd hear about them first in Bangalore."[43]

Learning in the Indian software industry occurs primarily within the firm and through long-distance relationships rather than through interactions with local customers, suppliers, or institutions such as universities and research laboratories. In India the public and private technology sectors, as well as industry and the universities, remain separate spheres, with limited professional or commercial interaction. Sridhar Mitta, CEO of Silicon Valley–based EnThink, a spin-off of Wipro, explains: "Bangalore has passed criticality in technical prowess but is still abysmally low on interaction. The culture of networking is better in Bangalore than in the rest of India, but nowhere near what exists in the Valley. Here a major part of that load is carried by multinationals which guard their secrets very jealously . . . If Silicon Valley scores 100 for the purpose of our present discussion, Bangalore scores 15."[44]

A domestic venture capital industry did emerge in India in the 1990s but, as in China, it differed significantly from that found in Silicon Valley. Much of Indian venture capital was provided by large public-sector funds or banks and by multilateral institutions. These organizations, like Union Trust of India Ventures or Small Industries Development Bank of India, lacked domain knowledge as well as experience in software or technology-related industries. The funds, typically run by risk-averse financiers concerned with tangible assets, were overwhelmingly late-stage investors in software services. In the words of A. V. Sridhar, former senior manager at Wipro, "The Indian venture capitalist will not take risks in new areas, as opposed to a risk-free definitive software services market, where he is assured of quick returns and profits."[45]

The supply of venture capital in India remains small by international standards, in part because of a multiplicity of conflicting, often cumbersome and anachronistic regulations and a variety of other forms of discrimination against the industry. In 1998 only 21 companies were registered with the Indian Venture Capital Industry Association, and they had approximately $700 million available for investment—as compared to Israel's 100 VC companies with $4 billion investable funds in 1999, and Taiwan's 110 funds with $1.32 billion in investments.

A few domestic technology start-ups survived in Bangalore during the 1990s, but none experienced the accelerated successes of start-ups in Taiwan or Israel at the time, even though most were started by returnees from the United States.[46] Silicon Automation Systems (SAS) was started in Silicon Valley in 1989 by four U.S.-educated Indian engineers. They bootstrapped the venture and a few years later moved part of the operation to Bangalore while seeking to maintain a strong orientation toward research and software and hardware design (as opposed to software services). For close to a decade SAS remained a small, privately held firm with about 300 employees and $10 million in exports. In 1999, Intel Capital made an investment in SAS and changed its name to Sasken Communication Technologies. The Intel relationship marked a turning point because it not only provided SAS with cash but also exposed the company to new customers, broadened its strategic vision, and provided opportunities for technical collaboration. Sasken is now one of the leading IP providers in India and employs more than 2,000 people.

Encore Software, another technologically sophisticated Bangalore company with a long history, designs and licenses DSP technology and embedded solutions as well as providing software engineering services in the same fields. Its customer list includes leading systems firms from around the world, among them Lucent, Huawei, D-Link, Infineon, Mitsubishi, Texas Instruments, and Analog Devices.

The firm was founded in 1990 as NCore Technology by Stanford graduate Vinay Deshpande, whose goal on returning home from the United States was to rebuild India's computer hardware manufacturing infrastructure. His first business, PSI Data Systems, became one of India's largest producers of PC-compatible computers before merging with the French company Groupe Bull. In 1995 NCore announced the "Yantra," a multimedia sub-notebook weighing 2.1 kg—the lightest in the world in its class at the time, and widely recognized as an innovative product. However, the firm's inability to finance long credit periods, its limited marketing resources, and continuing difficulties from customs and export-import laws ultimately undermined the effort. Deshpande reports: "There were customers who were willing to buy 20,000 customized machines annually from us. That's not a big number for the big companies, but is a very big one for us. However, no bank or financial institution in India would give us loans using the finished goods as security. They wanted collateral in the form of immovable assets or shares. They wanted credit of about $15 million. An IBM or Compaq would find no problem in giving this much credit, but we just did not have this kind of money."[47]

A breakthrough product like the Yantra would likely have received state funding or at least a loan without collateral from a development bank in an Asian country like Japan, Singapore or Taiwan, or Korea at the time. Sridhar Mitta, Chief Technology Officer of the Wipro Infotech Group, observed: "The product is an excellent one and something few Indian companies have attempted. Given that infrastructure in India is not available, it's all the more difficult. The idea from a design perspective is a very good one, but NCore did not have the marketing resources, the marketing experience." Mitta added that at the time the Yantra was developed, there was no comparable product on the market. "But the value of a design is zero in six months."[48] NCore eventually made a deal with Taisei Engineering to sell the machines through direct mail in Japan. A year later, however, Taisei decided to retain the computer for its in-house needs.

The firm continued to survive by licensing its technology, and in 1999 was merged into Encore Software Ltd. Deshpande continued to pursue his vision of building India's computer manufacturing capabilities. Encore collaborated with faculty and students at the Indian Institute of Science (IIsC) in Bangalore to develop the under-$200 Simputer (or simple inexpensive mobile computer), the first hand-held Internet device designed entirely in India and intended for the low-income Indian population. The Simputer gained international recognition as one of the first efforts to design technology for the poor, rural populations in developing countries.[49] In 2001 Vinay Deshpande was named one of the 100 Technology Pioneers by the World Economic Forum, and a year later the Simputer Project received the First Dewang Mehta Award for Innovation in Information Technology from India's Ministry of Communications & IT.

Growth and Constraint

The boom in the U.S. technology sector in the late 1990s had a mixture of consequences for India. First, labor shortages in places like Silicon Valley resulted in an increase in the quota for temporary visas granted on the basis of skill. In 2000 alone, 124,697 Indian nationals gained H1B status, representing 48 percent of the total U.S. visas granted that year. The next largest country in this category was China, accounting for only 9 percent of H1B recipients (22,570). Desai estimates that at least half a million Indian programmers received U.S. visas (of all sorts) between 1996 and 2001.[50]

Shortages of technical skill in the United States also contributed to the founding of cross-regional start-ups between Silicon Valley and India. Rakesh Mathur, who worked for Intel in Silicon Valley for many years before starting three successive successful technology companies—Armedia, Junglee, and Stratify—explains: "The key constraint to starting a business in Silicon Valley in the late 1990s was the shortage of software developers. I realized that I could go to India. All three of my start-ups had design centers in Bangalore but were registered as American technology companies."[51] Mathur made six trips to India in 1999, and in 2000 his firm Stratify established a 100-person operation in Bangalore.

The U.S. technology boom had other consequences for the entrepreneurial environment in India. When Infosys became the first Indian company to list shares on NASDAQ in 1999, it generated about a hundred "paper" or "dollar" millionaires and demonstrated the value of stock options locally. Similarly, Rajesh Jain became a local hero in 1999 when he sold his Mumbai-based company, IndiaWorld, to Satyam Infoways (India's first private Internet service provider) for $115 million (U.S. dollars). Jain held a degree from IIT Bombay and a master's degree from Columbia University; he worked in telecommunications in New York for several years before returning to India in 1995 to start IndiaWorld. Jain is now managing director of Netcore Solutions, an enterprise software firm focused on messaging, collaboration, and security software, and is one of India's first repeat entrepreneurs.

The fast-rising stock market valuations of domestic software firms like Satyam and Infosys attracted a new generation of venture capitalists to India. A group of $50 million to $100 million funds, typically from established firms such as Walden International and E-Ventures (Softbank), targeted early-stage investments. Corporate venture capital also became active in Indian technology regions. Intel Capital, for example, made a commitment to invest $100 million in Indian IT start-ups during 2000. Traditional sectors that once shunned the software industry became interested in investing, and fund managers began to tap India's high-net-worth individuals and family firms for capital. Total VC investments in India peaked at $1.2 billion in 2000, then fell off substantially for the next three years, but in 2004 rebounded to $1.1 billion.[52]

Permanent returnees to India from Silicon Valley remained few and far between through the 1990s, but the professional and personal networks linking Indians in Silicon Valley to family members, friends, and colleagues at home, combined with access to low-cost travel and communications, generated an unprecedented rate of information exchange between the United States and India. A Silicon Valley engineer commented:

> I go back to India two or three times a year because of my work. There are parts of India where you take a train and go over there and they don't even have a rickshaw or a cab to take you to your destination. You have to walk. But everybody in the small town knows exactly what the job situation is in Silicon Valley. They know the H1B quota level, when it is filled, when it is open again. They know exactly what kinds of skills are required in Silicon Valley, not even in California, just Silicon Valley.[53]

Web sites like *www.return2india.com,* Non-Resident Indians Online *(www.nriol.com),* and *www.siliconindia.com* became increasingly popular among U.S.-educated Indians. They traveled home frequently during the 1990s, often returning to traditional arranged marriages, and many seemed torn between the familial and cultural pull to return home and what they regarded as superior professional and economic opportunities in the United States.

Overcoming Obstacles

By 2000 the successes of Indian entrepreneurs in Silicon Valley were widely recognized in India. At least a dozen technology companies with Indian founders had been publicly listed or acquired, including Juniper Networks, Ramp Networks, Cyber Media, Exodus, WebEx, Selectica, and Tibco Software. U.S. technology entrepreneurs who visited India were treated like media stars, and meetings of the newly founded TiE branches in Bangalore and Mumbai drew hundreds of people.

Successful Indian entrepreneurs from the United States also invested actively in Indian technology start-ups, both directly as angel investors and indirectly through commitments to venture capital funds, and they demonstrated their loyalty to their alma maters—universities like the Indian Institutes of Technology—with multimillion-dollar contributions. The flow of returnees increased significantly, and the non-resident Indian (NRI) community, which had formerly been both resented and scorned, was increasingly welcomed in India and accorded growing public status and privileges.

The boom was reflected in already congested Bangalore. Infosys received a million applications for the 10,000 jobs it advertised in the first three months of 2004. Wipro hired more than 9,000 workers in the same period. In the words of Werner Geortz, a Wipro vice president who visited Bangalore in early 2004, "It's like a gold rush town here. You can't get a hotel room. There are headhunters outside the office complexes, waiting to jump [chip] designers when they leave work. It looks like nothing so much as Silicon Valley in 1999 [just before the bubble burst]."[54]

Silicon Valley's most successful Indian entrepreneurs, recognizing the obstacles to entrepreneurship in India, began to play an active role in policy. In the late 1990s Kanwal Rekhi, former CTO of Novell and TiE founder, gave a series of high-profile speeches and interviews in India in which he challenged the government and people to adopt modernizing reforms. He also met regularly with senior Indian policymakers. T. N. Srinivasan gives a large share of the credit to Rekhi for the "breathtaking scope of India's telecommunications reform and its success."[55] Rekhi's lobbying and public attacks on the deeply entrenched Department of Telecommunications (DOT) were crucial to breaking its monopoly, thus dramatically lowering telecommunications costs and widening access. Rekhi coined the Hindi slogan "Curb the DOT and Save the Nation!" Many of his proposals were implemented, including the liberalization and privatization of the domestic long-distance market as well as the later opening up of the international long-distance sector to private competitors.

Indian policymakers turned to the non-resident Indian (NRI) community in Silicon Valley for policy advice in other areas, leading one journalist to quip that the "brain trust" was replacing the "brain drain." In 1999 the Securities and Exchange Board of India convened the Committee on Venture Capital, led by successful SV entrepreneur K. B. Chandrasekhar, with the task of recommending steps to promote venture capital in India. The committee's report developed a policy agenda for the industry that appears to have contributed to the subsequent expansion of venture capital in India.[56] Silicon Valley's NRIs also spearheaded a plan to attract foreign direct investment to India and presented it to the government in 2005. It was well received, and several of its recommendations were implemented.

Silicon Valley entrepreneurs also financed policy-oriented research and conferences at Stanford University, designed to bring together academic researchers, top Indian politicians, and government officials to debate policy reform. Some individuals in the same network, recognizing that one of the key areas requiring policy reform in India is the state level (where democratically elected governments have less power and autonomy than do provinces in China), started to create business advisory groups for the Indian states based in the areas where network members were born.

INDIA'S CROSS-BORDER START-UPS

While firms like Ramp Networks represent the standard model for building a cross-regional company, starting in the United States and tapping talent in India, Draper International pioneered the reverse strategy. In 1997 the venture capital firm recruited A. V. Sridhar, a senior manager at Wipro who had never worked outside of India, to move to Silicon Valley to start a company in the data mining field. Sridhar quickly identified a marketing team, including Sanjay Anandaram, who was already working for Wipro in Silicon Valley, and senior managers with experience at Oracle India and IBM Research. The start-up, Neta Inc., which developed Internet personalization software, was acquired two years later by the Internet portal Infoseek.

Sridhar explains why he moved to Silicon Valley to start Neta: "To create a successful company, one has to be real[ly] close to the market. One has to be in a place which supports the creation of new technologies as a daily affair."[57] The regular interaction that comes from proximity to customers, and from trying to diagnose and solve problems, he says, helps generate original product ideas and new business models: "Silicon Valley is a risk-taking environment, whereas the Indian software model is largely based on cost-effective and efficient delivery of services."

This insight underlies the growing popularity of the cross-border start-up. There is an increasing trend for new ideas and business models to emerge from Silicon Valley, where engineering problems are defined, and then to be executed in India. Anandaram, cofounder of Neta, comments:

> The future is all about international companies irrespective of their size and stage. . . . As the U.S.-India technology corridor grows in the coming years, the effect of international processes, talent, and capital on economies like India will be huge. Technology companies will emerge out of India that need to be married into the U.S. start-up ecosystem simply because the required ecosystem (e.g., local capital, management talent, local market, legal framework) will take a while to develop inside India.[58]

This logic motivated a new generation of venture capital investors with experience in both India and Silicon Valley. Most venture capital in India is focused on safer, later-stage investments in software services or BPO firms. However, venture capital firms with Indian fund managers, including WestBridge Capital Partners and Artiman Ventures, have targeted early-stage U.S.-India crossover firms. The economic logic for this approach appears compelling: a company that would need $10–15 million in its early-stage funding in Silicon Valley might hire a comparable engineering workforce in India for only $2–3 million.

In July 2000, three veterans of the Indian IT industry established the JumpStartUp Venture Fund in Bangalore: Kiran Nadkarni, who had fourteen years of experience with venture capital in India including the position of Draper International partner in Bangalore; K. Ganapathy Subramanian, who came from ICICI Venture, a leading venture capital firm in India; and Sanjay Anandaram, who had worked in both Bangalore and Silicon Valley. The $45 million fund, targeted at early-stage IT start-ups, had funding from both institutional investors (including the Silicon Valley Bank) and individuals such as Bill Draper and successful Indian executives in the United States.

A company called July Systems exemplifies the new cross-border technology start-up: it combines the Silicon Valley ecosystem with India's low-cost technological capabilities. It is unique because it targets the fast-growing technology markets in Asia and India as well as those in the United States and Europe. The firm, started by Ashok Narasimhan, the founding president of Wipro Infotech and Wipro Systems (and one of India's most highly regarded technology managers), and second-time Indian entrepreneur Rajesh Reddy, is developing products and solutions that help mobile telecom operators monetize data services. First-round investors in July Systems included a Silicon Valley venture capital firm along with three regional VC firms, two that provided links to India (JumpStartUp and WestBridge Capital Partners) and one with strong Asia connections (Acer Technology Ventures).

July Systems has a fifteen-member business development and marketing team in its Silicon Valley headquarters, and forty-five employees in its Global Product Center in Bangalore (including about a dozen Indians with experience in Silicon Valley) who are responsible for product development and engineering as well as product marketing, program management, sales support, customer support, and finance. The firm also has sales offices in Asia and Europe. July Systems is one of the many high-profile entrepreneurial spin-offs from India's Wipro Technologies, which has become a managerial training ground for Bangalore, playing a spawning role similar to that of Fairchild in Silicon Valley and Acer in Taiwan.

Despite growing interest from international investors, India's venture capital industry—like that of China—is still constrained by underdeveloped capital markets and extensive financial regulations. In 2000 a former head of R&D at Wipro, Sridhar Mitta, established an incubator in Bangalore called Enterprise for Entrepreneurs, or e4e Labs, committed to investing $300 million in ten to twelve companies over three years. The firm is based in Silicon Valley because, as Mitta notes, there is not yet a clear exit option for venture capital investments in India: "The Indian IPO is not proven yet, even on the services front. Big successes have still not come in to make the way easier for venture capital newcomer entry."[59]

Back to the Valley

In 2002 the early-stage venture firm JumpStartUp moved its headquarters from Bangalore to Santa Clara, California, reflecting a shift in its investment strategy. Like Draper before, it abandoned the idea of an India-focused fund for stand-alone start-ups and instead focused on "U.S.-India cross-border investments." The partners realized that their small fund was not sufficient to support early-stage start-ups from the ground up in a recessionary environment when outside investors were reluctant to contribute to risky new Indian companies.

The firm's new strategy recognized as well that Silicon Valley's cash-strapped start-ups were being pressured by their investors to set up engineering centers in low-cost locations like India. JumpStartUp now envisions a role as co-investor with established VC firms in order to help portfolio companies set up engineering teams as well as even design, deployment, and support functions in India. In the words of Kiran Nadkarni, "It is very hard for companies started by non-Indians to think of working out of India, unless they have done it in the past. Unless the founders have [the intention] to do things out of India . . . When a start-up decides to establish a development center in India, invariably you will see that one of the founders is an Indian."[60]

Silicon Valley Bank executive Ash Lilani estimates that 10 percent of his customers, or some one thousand venture-backed companies, were doing some development in India in 2004, and he anticipated that the proportion would grow to 25 percent in a couple of years: "There was a time when start-ups didn't think globally about anything. Now entrepreneurs are thinking at the formation stage about what is the best model for the company. You see the cross-border model from day one."[61]

Up the Value Chain: IC Design

The growth of IC design collaborations between India and Silicon Valley represents a new, higher-value-added trend. Companies like Intel, Texas Instruments, Analog Devices, and Cypress Semiconductor have operated sophisticated IC design operations in Bangalore and other Indian cities since the early 1990s, but they remained relatively low-profile. A series of highly publicized acquisitions by Silicon Valley firms of India-based semiconductor design companies since 2000 have fueled interest in the sector: Broadcom acquired Armedia, Intel acquired ThinkIT Systems, and Cypress Semiconductor acquired Arcus. Now a growing number of start-ups are relying on Indian IC design capabilities. Insilica, Open Silicon, Ample Communications, and Atrenta have all been launched since 2000 by experienced Indian engineers who remain in Silicon Valley while leveraging design talent in their home country.

Silicon Valley's best-known Indian chip designer, Vinod Dham, and partner Tushar Dave have started a venture capital firm, New Path Ventures, to fund cross-border semiconductor start-ups that will combine U.S. headquarters, Indian development talent, and Asian (Chinese or Taiwanese) foundries to establish the design in silicon. Dham formerly headed the Intel team that developed the Pentium chip and later started Silicon Spice, which was acquired by Broadcom. He explains:

> Software service models involve low investment and fairly steady revenues. We need to break this mindset and move to the next level in the value chain. Chip design should be the natural progression for Indian engineers to take up. Today, if you heave a rock in Bangalore, you are bound to hit a UNIX or Oracle software engineer. Such quality resources should not be wasted in mundane jobs, and should be challenged to innovate, develop, and deliver superior products that come with the pride of ownership. . . . We will see a lot more engineers who are "chip heads" and not "software geeks."[62]

By 2005 India was home to between 6,000 and 10,000 IC designers, a fraction of the nation's 650,000 software engineers.[63] However, these designers earned significantly more money and often created their own IP rather than simply providing services.

Ittiam Systems is one of Bangalore's most successful recent venture-funded start-ups, designing digital signal processor (DSP) software for applications such as streaming video, portable media players, and digital still cameras. The firm was started in 2001 by a team of experienced senior managers from Texas Instruments (India) including the managing director, Srini Rajam, now CEO of Ittiam. Rajam notes that the firm avoided a service model by licensing the intellectual property (IP) to clients such as Sony and VEO and Texas Instruments. In the words of Rajam,

> The IP model allows you to leverage India's strength, which is its intellectual capabilities. [An IP model lets] you work on creating new algorithms, designs, and applications and at the same time you avoid the logistical headache, which is supply chain, distribution, logistics, and so on.[64]

Rajam also notes that "very few companies can afford to invest the time and effort to stay at the forefront of so many technologies [such as video, telecom, imaging, computing, and audio]. By licensing our design innovations, they can cut the time it takes to launch a product by 9 to 12 months."[65] In 2004 Ittiam Systems was named the most preferred supplier of DSP-based IP by a U.S. market research company, Forward Concepts.

At the same time that entrepreneurs were pioneering IC design, a group of leading software companies in Bangalore recognized the need for collective effort and created a nonprofit IP Exchange Forum to help India achieve leadership as an electronic product design hub. The group's goal was to emulate the Taiwanese experience in personal computers by creating an ecosystem for IP that encourages the development of technical talent, provides methodologies for evaluation, contributes to standards creation, provides market data and technology trends, supports collaboration in development and marketing, reduces duplication in IP development, and enhances the credibility of India in electronic product design.[66]

While the shortage of qualified talent in India is the dominant constraint on the expansion of the domestic IC design industry, the lack of semiconductor fabrication facilities may limit its growth in the future. The only viable fab, Semiconductor Complex Ltd. (SCL), is a government-run business located in northern India (Chandigarh). However, the Indian government has not been willing or able to attract the massive capital investment needed to upgrade the SCL facility. Many industry experts believe it is not worthwhile in an era when sophisticated foundry capacity is available in Taiwan and China—especially because India's backward infrastructure, unreliable power, and limited water supply all pose major challenges to semiconductor manufacturing. But since successful IC design depends on an understanding of process technology, fabrication, and testing, the creation of a semiconductor ecosystem may require at least a couple of fabs as well as investment in university training and research.

Travel and Trouble

The Indian counterparts of the Chinese who appear to spend their lives on airplanes have emerged since 2000 from the new cross-border venture capital firms and their cross-border start-ups. These investors and engineers travel for 25 hours, or halfway around the globe, for meetings. They face a whole new set of challenges associated with the management of start-ups that are based in distant time zones, have distributed product development, and are characterized by cultural differences. Naveed Sherwani, CEO of the ASIC design start-up Open Silicon (whose staff is split evenly between Bangalore and Silicon Valley), says that many people underestimate the differences in work culture. Indian employees, for example, are not accustomed to open discussion with their bosses. "If an employee in North America says 'I will do it by 5 p.m.' it will be done by 5 p.m. Or they will tell you they can't do it. In India, out of respect, they say they'll do it, but it may or may not actually get done."[67]

Venture capitalist Ravi Chiruvolu similarly explains that the process of setting up operations in India, even in 2003, is not at all simple:

> Talent is not as cheap as we had thought. Basic office infrastructure is not quite as easy to obtain. Business partnerships are fraught with government and political loyalties and traditions. Even office leases can become complex entanglements.
>
> And that's just speaking from our own experience, not just as venture capitalists but as ex-pat Indian venture capitalists, investors who literally speak the language and understand the culture of Bangalore's "Silicon Valley" better than most.[68]

Other firms have found the problems of collaboration over great distances to be insurmountable. Sonim Technologies, a wireless technology company, started with most of its engineers in Bangalore but after three years decided to move them back to Silicon Valley because the distance was slowing down product development. Sonim engineers needed daily contact with U.S.-based cell phone partners, but the 12-hour time difference made this virtually impossible. According to Sonim's CTO, "We needed to get the product out faster. . . . We had to have everyone under one roof and much closer to the partners."[69] Firms like Sonim have learned that the range of things that can be communicated and carried out long distance is limited; that the costs of communication and travel are often higher than anticipated, as are the costs of the missed deadlines that result from poor communications; and that success requires both discipline and work that is clearly partitioned. Some firms now face turnover rates in excess of 50 percent per year in India.

The availability of bandwidth and sophisticated software tools promises to reduce these challenges. Once two engineers trust each other, through experience or face-to-face interaction, it appears to be possible to create the equivalent of a "virtual water cooler." The COO of MindTree Consulting, Subroto Bagchi, notes that in the 1990s it appeared that customer interaction, collaboration, and design all had to be done exclusively in the United States. However, that too appears to be changing: "If earlier we looked at India for just development or maintenance work, now we are able to look at co-development and co-architecting."[70]

Institutional change takes time, however, and the resignation of Vivek Paul from his position as vice chairman of Wipro in 2005 was a reminder of the lasting legacy of India's experiment with economic autarky. While Paul remained discreet about his reasons for leaving, observers believed that Wipro chairman Azeem Premji, who owns over 80 percent of the company and still runs it like a family-owned business, was reserving the top leadership positions in the company for his sons, who are studying at business school in the United States. When asked, Paul admitted that "Indian industry is still living in the afterglow of the licence-permit raj. Indian promoters and managers have grown up in a cocooned world. There is no cocoon in the real world . . . In the U.S. they will do whatever it takes to get the best person."[71]

WHAT NEXT?

While Bangalore and its smaller counterparts continue to boom, they are at risk of becoming technology enclaves. The IT-related industries are far more heavily concentrated in these areas than elsewhere in India, but they employ at best a million people, a minuscule proportion of the total 410 million employees in India. India remains overwhelmingly a rural society, with 72 percent of the population residing in rural areas, and the economy is primarily dependent upon agriculture, which employs 58 percent of the workforce. Even in Karnataka the IT sector provides only 0.7 percent of all jobs, and they are heavily concentrated in the capital city of Bangalore.

Labor shortages and rising wages for experienced workers are now a serious problem in regions like Mumbai and Bangalore, where turnover is high and poaching of employees is common. Salaries for middle- and higher-level jobs have continued to rise as growing numbers of Indian companies and multinationals compete for a limited base of experienced engineers. As a result, the wage gap with the West is diminishing quickly at the higher levels, despite the abundance of fresh graduates and the absence of upward wage pressure on lower-skill, entry-level jobs.

The ecosystem for technology start-ups in Bangalore and other regions in India has improved over the past decade but, as in China, local start-ups remain distant from customers and lack the interaction and learning that characterize a local technical community. This is a result of the export orientation of local producers and their limited differentiation, as well as the traditional secrecy and hierarchy of older Indian businesses. The development of an entrepreneurial culture will require collaboration between firms, universities or research institutes, financial institutions, and customers and suppliers—a situation that is still rare in India today.

There are some positive signs. The long-standing separation in India between the ivory-tower university researchers and industry is beginning to break down. Strand Genomics in Bangalore, a life-sciences informatics company that uses data mining and analysis for drug discovery and development, was started in 2001 by a team of returnees from the United States, including three computer science faculty members from the Indian Institute of Science, with funding from WestBridge Capital. With faculty mindsets changing and students increasingly willing to assume risk, several of the IITs have spawned promising technology start-ups, including IIT-Mumbai, IIT-Chennai, and IIT-Delhi.

Collaborations between Indian IT companies and domestic companies with domain knowledge could become an important source of indigenous innovation. The largest Indian firms have been reluctant to serve the domestic market because the margins are small compared to those from exporting services. In addition, important sectors of the Indian economy from retail and health care to real estate and insurance remain closed and protected—hence less receptive to the opportunities provided by IT. Nonetheless, the domestic market could provide important opportunities for Indian entrepreneurs to contribute to problem solving and the development of intellectual property that they might eventually export.

A recently announced product is the Mobilis computer, a Linux-based mobile desktop computer that is lighter than currently available products and affordable to a wide range of the Indian population, with a price tag starting at 10,000 rupees (approximately $250). The computer is not only cheap but runs on batteries, eliminating the need for electricity, which is often unavailable in India and other developing economies. It is targeted at the needs of the Indian market, with built-in local language support and basic software applications. The Mobilis also represents an unprecedented collaboration in India, the New Millenium Indian Technology Leadership Initiative, which is coordinated by the Council for Scientific and Industrial Research (a government agency) and Encore Software Ltd. Encore plans to work with the independent software community to develop open-source applications for various areas ranging from governance to telemedicine.[72]

The evolution of political institutions matters tremendously in both China and India. India is formally more democratic, but in practice its central and local governments are more bureaucratic and less responsive than China's. Provincial governments in China have the authority and the incentive to aggressively promote local economic development, and most have done so successfully. State governments in India, however, lack resources to make necessary investments in development, and they are often politically gridlocked when it comes to reform. Finally, while both countries face corruption at all levels, China has made a high-level commitment to analyzing and overcoming obstacles to development. Such an effort is less evident in India.

Both China and India have trouble financing entrepreneurship. India's financial system is more effective than that of China: private-sector firms, particularly small- and medium-sized firms, have significantly better access to capital through the Indian banking system and the Mumbai stock exchange. However, India's domestic savings rate compares poorly, at 22.5 percent of GDP in 2004 compared to 47.6 percent in China, which will continue to constrain much-needed investments in infrastructure and public services. Nor will foreign investment make a meaningful impact. In 2004 India received only $5.3 billion in foreign direct investment, compared to

China's $60 billion. The fiscal deficits of India's central and state governments remain large as well, which suggests that financial reform will be required if India is to raise the capital needed to support sustained growth.[73]

India should look to China for a model in developing its infrastructure, which has been neglected in Bangalore and other technology regions despite the tremendous population growth brought on by the technology boom. Roads are heavily potholed and appear to be permanently congested; the construction of an international airport in Bangalore has been delayed repeatedly; power outages and water shortages are a regular occurrence. In 2003 Wipro founder and head Azim Premji announced that the firm would no longer expand its business in Bangalore: "It is embarrassing when there are four power cuts during one hour of discussion with clients."[74] While not all Indian regions face the same degree of infrastructure problems, none begin to approach the reliability found in the emerging technology cities in China. Even Internet usage remains low in India, at 3.6 percent of the country's more than one billion people, compared to China's 8 percent.[75]

Investments in education and research are more important than ever for India today—yet they too remain stalled. Although India produces large numbers of engineers each year, their quality is uneven. Most higher education in India is still geared toward producing "babus" (a pejorative term that refers to clerks and petty bureaucrats). Only a few regional engineering colleges, along with the Indian Institutes of Technology and the Indian Institute of Science, are training world-class engineers; and the country produces a very small number of doctorates. The public university system, which is politicized and often corrupt, faces a shortage of high-quality teaching staff and severely limited resources.[76]

The deepest challenge for India, however, will be to confront the stratification that grows out of its caste system. IT professionals come predominantly from Brahmin and other upper castes. Although employers report they are hiring on the basis of merit rather than caste, the structure of educational and professional opportunity in India is largely self-perpetuating: those who have access to private schools have access to the elite universities—and they are overwhelmingly from urban, upper-caste or Jain backgrounds.[77] Poor lower-caste and rural residents are consigned to inferior public schools and rarely qualify for higher education. While some may get basic computer training, they remain in lower-level jobs with limited prospects for social mobility. The inequality of educational opportunity in India is typical of developing economies in Latin America and elsewhere; but this is not true of China, where, despite significant income inequality, widespread access to standardized education has preserved social mobility.[78]

The need in India for improved access to, and investment in, education and other public services, ranging from health care to infrastructure, is immediate. More than a third of the Indian population today is under the age of 15, and more than half is under 25.[79] The future of India in this century rests with this youth population boom—this half of the population could provide the skill, energy, and vision needed to transform the nation into a dynamic and developed economy, or they could grow up in an increasingly unequal society without employment opportunities, creating enormous social pressures. The challenge of overcoming political patronage and gridlock to undertake serious economic and political reform is imperative. The IT industry, while small in terms of employment, could play a far-reaching leadership role in this effort.

8 ❖ The Argonaut Advantage

The new Argonauts have redefined both Silicon Valley and its role in the global economy. The Valley is famous for its technological innovations, but it is the organizational innovations that provided a regional advantage in the twentieth century and are a source of the Argonaut advantage today. Entrepreneurial communities, first from Israel and Taiwan and later from China and India, have transferred Silicon Valley's system of open networks and decentralized experimentation to their home countries while at the same time remaining close to its fast-changing markets and technology. By promoting the development of local capabilities in Tel Aviv, Hsinchu, Shanghai, Bangalore, and other technology clusters while also collaborating with entrepreneurs and investors in Silicon Valley, the new Argonauts have initiated a process of reciprocal regional transformation that is shifting the global balance of economic and technological resources. Silicon Valley, once the uncontested technology leader, is now integrated into a dynamic network of specialized and complementary regional economies.

These new technology regions are not replicas of Silicon Valley, nor are they becoming new Silicon Valleys. We have seen that in each of these regions the new Argonauts face obstacles to technical and entrepreneurial development rooted in the limitations of local financial, regulatory, and legal institutions, capital markets, management models, politics, demographics, foreign relations, and so on. Even as the returnees seek to use their experience in Silicon Valley to reshape these institutions, distinctive regional and national histories ensure that the identities and technology trajectories of these regions are unlikely to converge.

The new Argonauts are, however, building technological and entrepreneurial capabilities in distant regions. They serve as role models, mentors, partners, and investors for entrepreneurs in their home countries—just as earlier generations of immigrants did for more recent arrivals in Silicon Valley. They engage policymakers on policies to improve the local environment for entrepreneurship; they emphatically reject the familial, opaque, and frequently corrupt business practices that dominate in many developing economies. In short, they transfer institutional know-how as well as technology, capital, and contacts to their home countries. They do not do this selflessly, though there is an element of personal commitment for many. Most are motivated by professional challenges as well as economic opportunity. Emigrants usually return home only when they come to believe that their capabilities will be more highly valued in their native environment than in the United States.

FROM HIERARCHIES AND MARKETS TO OPEN NETWORKS

Researchers who study late economic development typically credit multinational corporations or a state committed to economic development with mobilizing the technology, skill, and capital required for poor economies to catch up. However, the cases examined in this book suggest that the process has been far less predictable and more bottom-up than these models assume. Governments have invested in education, research, and infrastructure in each of these regions—

in some cases with the active guidance of the returnees. Multinational companies have served as role models, competitors, and partners for local enterprises. But governments and multinationals alone could not have achieved the combination of local capability building and global integration that characterizes modern technology regions. The Argonauts have provided the essential mix of local knowledge and global connections required to initiate and motivate the experimentation and co-development that are transforming producers in once-peripheral locations into technology leaders.

Mainstream economists have argued that digital technology empowers individuals and supports a return to free markets. In this view, the hierarchies of the vertically integrated corporation are no longer needed to coordinate production and distribution because computers and broadband networks, along with technical standards (or design rules) governing the interfaces between stages of production, have dramatically reduced the information costs of coordination. Firms and individuals, in this account, can tap low-cost skill and other resources anywhere in the world instantaneously; and global supply chains can be coordinated through arm's-length exchanges between producers of standard components.

Managerial hierarchies are indeed being flattened and transformed as enterprises reap the benefit of access to real-time information, lower cost, and high-quality alternatives to internal sources of components and services. Information technology producers have been leaders in the adoption of open, Web-based platforms, process standards, and applications for coordination of long-distance exchanges of information and goods. This mix of vertical disintegration and virtual integration provides greater speed and responsiveness than the hierarchical corporation—and has proven superior in today's uncertain environment.

The design of new technology systems and components, however, involves more intense communications and interaction than design rules, computers, and bandwidth alone can provide. The Argonauts have contributed to the development of new non-market mechanisms of coordination as well. Their embrace of rigorous quality standards and benchmarking processes such as ISO and SEI CMM, initially as a means for peripheral producers to establish their credibility and reputation in global markets, has promoted continuous incremental product, process, and cost improvements. These small but continual innovations cumulatively contribute to shorter product development times and improved product quality in both IT manufacturing and software development.

New technology products like PDAs and laptop computers combine components, services, and intellectual property from around the world, but these systems and their sophisticated components are rarely defined entirely in advance. They are designed through a process of simultaneous engineering in which specialists collaborate to continually innovate and reduce costs.[1] Information systems now facilitate this iterative co-design, even at great distances, by accelerating the formalization of tacit knowledge and facilitating transparency.

Nevertheless, most new or complex products still require periodic face-to-face collaboration by technical leaders and product managers.[2] The new Argonauts retain an advantage in these long-distance collaborations because their shared language, culture, and professional and educational experiences help them avoid the miscommunication, cultural misunderstandings, and conflicting expectations that frequently plague long-distance work.[3]

The most successful enterprises in the new regions have achieved prominent roles in global supply chains. Well-known firms like TSMC and Infosys, as well as lesser-known but equally profitable firms like Ittiam and UTStarcom, regularly innovate through decentralized and disciplined collaboration with trusted partners. Trust is typically the outcome of this sort of successful collaboration, not a starting point. These companies benefit from scale economies in producing particular components or processes, but these advantages are typically short-lived. Faced with continuous pressure to rapidly build upon unique strengths in order to remain at the leading edge, producers of even mundane inputs seek to differentiate the quality or performance of their products. Although some technology components are standardized and sold on the basis of price, few remain in this category for long. As capability building and innovation increasingly occur at all levels of the supply chain, hierarchical control is replaced with horizontal interfirm collaborations that often blur the boundaries of the enterprise, and even the industry.

CROSS-REGIONAL COLLABORATION AND CAPABILITIES

By the early 1990s, it was clear that Silicon Valley no longer enjoyed a monopoly over the talent and know-how needed to innovate in IT production. Though largely unobserved during the Internet boom, producers in many parts of the world began to exploit the openness of global supply chains by introducing new components, software, or systems or by improving the quality, performance, and cost-effectiveness of particular subsystems or components.

These new technology regions initially entered the global market by providing low-cost skill, but each region has developed distinctive specializations that allow it to compete on the basis of quality, efficiency, or innovation as well as cost. Taiwan was known in the 1980s for its cheap PC clones and components; today it is a leader in global logistics and design. China was known in the 1990s for "me-too" Internet ventures; now Chinese producers are poised to play a central role in the development of wireless technology. In the 1990s India was a provider of labor-intensive software coding and maintenance; today local companies are managing complex consulting projects for global customers. Israel was a low-cost location for research and manufacturing in the 1980s; since then local entrepreneurs have pioneered sophisticated Internet and security technologies.

The new technology regions have inexorably become higher-wage, higher-cost locations within their national economies. Most face labor shortages at high skill levels, and yet they continue to spawn start-ups and attract subsidiaries of established firms. This is reminiscent of the continued clustering of electronics-related firms in Silicon Valley during the 1980s and 1990s; these areas now enjoy a formidable regional advantage that compensates for many of these costs. Silicon Valley producers no longer view locating in or sourcing from India or China as an efficient way to reduce costs; instead they frequently argue that the reason to do work in those locations is to gain access to local talent.

As we have seen, U.S. technology producers now look to their counterparts in Taiwan, China, India, and Israel not simply for low-level implementation but increasingly for co-development and co-architecting of products and components. More and more frequently, firms in new technology regions are partnering with one another as well as with firms from Silicon Valley, as when a Taiwanese semiconductor firm invests in Israeli start-ups specialized in digital speech processing chips, or when an Israeli company contributes intellectual property components to a chip design firm based in India. These collaborations deepen the capabilities of both partners and over time can support a process of reciprocal innovation in the respective regions.

As enterprises open their boundaries in new ways, the meaning of a company is being continually redefined. The proliferation of cross-border venture capitalists and start-ups underscores the advantages of combining the specialized and complementary capabilities of producers located in distant regions. It is the region, with its cluster of specialized producers and capabilities, and the networks that span these regions more than the individual firm that increasingly defines the contours of global production. In the near future, new systems architectures and technologies will continue to be developed in Silicon Valley because new product definition requires close interaction with customers.[4] The United States remains the world's leading market for technology products and services and also boasts the largest pool of experienced computer scientists, software engineers, and systems architects. None of the new technology regions has the depth of managerial and investor experience, or the depth and diversity of technological and research capabilities, found in Silicon Valley.

Technology markets are shifting quickly, however, with demand growing rapidly outside the United States. Whereas in 1985 the Asia Pacific region (excluding Japan) accounted for only 6 percent of semiconductor consumption, by 2000 it accounted for 25 percent, and this share is predicted to grow to 46 percent by 2010 (with China accounting for 16 percent). China and India are now the largest and fastest-growing markets for wireless technologies, and this is creating opportunities for local producers to design new architectures and products for their domestic markets. These markets are also likely to provide product and design inputs for the United States as well as for developing-economy markets. Text messaging on mobile phones, for example, originated in Asia and Europe but has become increasingly popular with American youth. Likewise, the fuel cells developed to provide a low-cost, long-lasting source of electricity for the Chinese and Indian markets could also change the economics of energy provision in the United States.

Today the West and Japan account for less than 15 percent of the world's population but produce more than half of global output. This situation will change decisively in the next fifty years, as customer demand from India and China comes to dominate global markets. Producers in other regions will surely develop sufficient technical skill and capabilities to participate in global networks as well. They will likely share with their predecessors a history of investment in education and research, as well as a willingness to experiment with institutional openness. Silicon Valley's role as the world's dominant technology center will continue to diminish. This does not mean that the region will decline, however. As long as local producers continue to reinvent themselves as they have done in the past, the region can flourish as one of many nodes in an open and widely distributed global network of specialized and complementary regional economies.

OBSTACLES TO REGIONAL DEVELOPMENT

Some of the largest immigrant groups in Silicon Valley have not built business or professional connections to their home countries. Most of the region's Iranian and Vietnamese immigrants, for example, are political refugees and thus not inclined to return to countries that lack the economic stability needed for technology investment or entrepreneurship. This applies in varying degrees to many of the developing economies that have technically skilled communities in the United States and at home, including Russia, parts of Eastern Europe, and Latin America. It is possible that urban areas like Saint Petersburg or Buenos Aires could become more attractive to returning entrepreneurs in the future as their economies develop, providing greater professional opportunities for returnees. To date, however, large parts of Africa and Latin America have yet to build the base of skill and the political and economic openness needed to become attractive environments for technology entrepreneurs.

Advanced industrial economies face different challenges and obstacles to building dynamic technology regions. Although they typically boast well-developed skill bases, research capabilities, and sophisticated infrastructure, technical skill is significantly more costly in these economies, particularly relative to recently developed economies like Taiwan and Israel. Financial and corporate structures in many industrialized economies discourage technology entrepreneurship as well. In Japan and France, for example, the privileged relationship of large corporations, banks, and the state limits the opportunities for outsiders (such as returning entrepreneurs) to raise capital or gain access to domestic markets. Europe's ongoing fascination with industrial champions suggests that policy there may continue to reinforce, rather than dismantle, these relationships.

Labor markets mirror this orientation. The technical elite in countries like France and Japan move automatically into high-status positions at the top of the large corporations or the civil service. They have little incentive to study or work abroad, and often face significant opportunity costs if they do. As a result, relatively few pursue graduate education in the United States, and those who do often return home directly after graduation. Those who end up in Silicon Valley for a period are not likely to gain access to capital, professional opportunities, or respect when they return home.

In many Asian countries government promotion of large-scale, capital-intensive investments during the 1970s and 1980s, either by domestic corporations (in Korea) or by multinationals (in Singapore), has similarly created environments that are inhospitable to entrepreneurship. This is confirmed by data on the sources of innovation. South Korea's *chaebol,* or large business groups, accounted for 81 percent of all U.S. patents earned in Korea in the 1990s, compared to only 3.5 percent earned by similar business groups in Taiwan.[5] Similarly, in South Korea the top fifty assignees accounted for 85 percent of all U.S. patents, with Samsung alone accounting for 30 percent, whereas Taiwan's top fifty assignees accounted for only 26 percent of U.S. patents. This decentralization of innovative capabilities was reflected in a substantially higher rate of patenting in the late 1990s, with Taiwan earning 17.7 patents per billion dollars of exports compared to Korea's 11.6.[6]

Since the 1970s, multinational corporations from the United States,

Japan, and Europe have aggressively expanded labor-intensive electronics assembly and manufacturing in low-wage locations. These investments extended the geographic reach of these traditional Chandlerian firms to places like Singapore, Malaysia, Scotland, and Ireland. They contributed to local mastery of high-volume electronics production, as well as to substantial improvements in local standards of living.[7] However, the hierarchical control over these subsidiaries typically discouraged the creation of autonomous technological capabilities. As a result, the leading recipients of foreign direct investment remain technological laggards. The rate of patenting, normalized by either population or exports in Singapore, Malaysia, Hong Kong, and Ireland since 1970, remains a small fraction of that observed in Taiwan and Israel.[8] Seven of the top ten patent recipients in Singapore, for example, were foreign multinationals or organizations; they accounted for 46 percent of all U.S. patents between 1970 and 2000.[9] Recent policy changes, such as public support of venture capital, have not been sufficient to transform domestic institutions or capital and labor markets. Most skilled workers in these places prefer stable, corporate employment; entrepreneurs have difficulty locating the managerial and technical expertise, as well as the market relationships, needed to build a new company.

The structure of Korea's and Japan's technology industries also poses problems for aspiring returnees. While Silicon Valley's Chinese and Indian entrepreneurs often grow businesses through mutually beneficial alliances with enterprises in their home countries, there are few such opportunities for Korean and Japanese entrepreneurs. The giant business groups (Korea's *chaebol* and Japan's *keiretsu*) that dominate these economies are unwilling to collaborate with small, distant enterprises.

The creation of a transnational community is a two-way process. While policymakers and planners can encourage cross-regional connections, they cannot create or substitute for transnational entrepreneurs and their decentralized networks. Government agencies from around the world regularly sponsor networking events for their expatriates in the Bay Area as a way to recruit returnee entrepreneurs and investments. However, when there are no entrepreneurial collaborators at home, these agencies can only provide incentives and information but not access to partners with an interest in jointly transforming the home environment. Governments cannot by themselves ensure the preconditions for return entrepreneurship; this is inherently a process of collaborative institution-building that takes both local knowledge and understanding of global technology markets and networks.

THE FUTURE OF SILICON VALLEY

Silicon Valley today is adapting to a competitive landscape in which other regions in the world are not simply sources of low-cost skill but innovators, potential partners, and competitors. The region has adapted to wrenching transformations in the past, including the 2001 Internet bubble, and may do so again. The adaptive capacity of a decentralized industrial system like Silicon Valley lies in its ability to rapidly redeploy resources to multiple, competing experiments, and its willingness and ability to learn from failure.

This does not mean that the regional adjustment process is automatic, or that the market will, by itself, provide an adequate response. Producers in Silicon Valley have benefited greatly from decades of postwar federal investments in technological research along with state-level commitments to primary, secondary, and postgraduate education. In the boom years it was easy to overlook or take for granted the contributions of these long-term investments to the economy's dynamism; in today's fast-paced global environment, however, it is more critical than ever to continually invest in regional advantage.

Policy discussions that blame outsourcing or globalization for domestic economic problems divert attention from challenges closer to home.[10] Silicon Valley has led the growth of the San Francisco Bay Area for three decades with consistent and robust productivity growth. In 2002, productivity in the Bay Area was double the U.S. average, with output per employee in key technology sectors of $63,400 compared to the nation's $32,700. The region's firms consistently attract one-third of the total venture capital invested in the country. They enjoy access to one of the world's most highly educated workforces, with 45 percent of residents in the 25- to 34-year-old age group holding a bachelor's degree or higher, compared to 26 percent nationally. The diversity and depth of managerial, technical, and professional know-how in the region remain unsurpassed.

The United States maintains a definitive technological lead over both India and China, which lack the technological skills as well as the university research capabilities found in most industrialized economies.[11] Both also face political and economic challenges that will take decades to overcome. Even the brain drain is not likely to end in the near future: many U.S.-educated Chinese and Indians working in Silicon Valley today are still reluctant to return home to work until they have obtained permanent residence in the United States. And while foreign student enrollment in American universities has fallen in recent years as a result of a more burdensome visa process and concerns about hostility to foreigners, the best students from China and India are still likely to prefer a U.S. graduate degree if it remains available and affordable.

The dwindling state and federal commitment to high-quality public education and research nonetheless threatens to undermine both Silicon Valley and the nation's technology future. If the foreign-born graduate students who once filled U.S. computer science and engineering programs continue to stay home, the nation's deficit of scientists and engineers will become more costly. If university researchers must compete for grants that are tied to short-term security objectives rather than longer-term and more open-ended basic research, our research capabilities and technology pipeline will atrophy. Silicon Valley's future—like that of its competitors—depends on ongoing investments in the physical and technical infrastructure, university research, and the skills of the American workforce.

The high and rising costs of doing business in Silicon Valley require that enterprises regularly adapt by pursuing higher and higher value-added activities. Local manufacturing became too costly in the 1980s, and by the 1990s many software engineering and development tasks were no longer viable. In the words of Eric Benhamou, chairman of 3Com and former CEO of Palm Computing:

> There's a whole range of tasks that 10 years ago would be performed only by sophisticated engineers in Silicon Valley that can now be performed by any one of 4 million Ph.D.s coming out of the Shanghai-Beijing region or the equivalent region around Bombay . . . jobs that used to be viewed as high intellectual capital are really moving out. This is not just support jobs, software QA jobs. You can actually do original development in Asia. . . .
>
> In the mid-90s, if you wanted to design a new protocol for the Internet, something that would become the eventual standard, this was the place to recruit the software engineers to do that. Today a networking equipment vendor in China . . . has people who know as much as anyone else who attends the IETF [Internet Engineering Task Force] meetings. Similarly, if today you want to build a 50,000 gate ASIC, chances are you could build this with an ASIC lab over there as well.[12]

Benhamou goes on to argue that Silicon Valley remains ideally suited to certain activities:

> If on the other hand, you try to figure out what is the truly innovative way by which consumers will manage the electronic equipment

> in their households in three years, after they've accumulated all these PVRs [personal video recorders], digital cameras, and Palms, chances are the breakthrough ideas will come from here. The way to connect these ideas to business and market will come from here.

Silicon Valley is still the best location in the world for the definition of new system architectures, high-level design, management of cross-cultural projects, and basic as well as first-generation engineering research. The challenge will be to maintain the concentration of customers, competitors, and partners; the experienced engineers, managers, and professional service providers; and the open social networks that foster the unanticipated cross-fertilizations of expertise and technology. As Benhamou puts it: "There's something that the Internet has not replaced and has not solved, and that's why Silicon Valley is so essential. There's nothing that can replace the chance-encounter, face-to-face discussion between business people from slightly different areas, from slightly different perspectives. You cannot convey passion and excitement even across the high quality of videoconferencing as you can here face-to-face. Often these breakthroughs occur in high-spirited meetings between passionate people. [If you] live 5,000 miles apart and never see each other, these meetings will not occur as often and they will not create the same sparks."

The new Argonauts will remain vital to the global economy as they pave the way for the development of new centers of technology entrepreneurship and new technology markets. They are most likely to favor regions that educate their youth, forgive failure and reward success, resist the impulse to protect yesterday's markets and jobs at the expense of tomorrow's, and welcome the openness, diversity, and initiative that have built Silicon Valley. Silicon Valley likewise has much to gain from participating in a widening network of more and more capable partners, both local and long-distance, that can provide new skills, insights, and resources to solving human problems.

APPENDIXES

NOTES

REFERENCES

ABBREVIATIONS

ACKNOWLEDGMENTS

INDEX

APPENDIX A

Immigrant Professional and Networking Associations, Silicon Valley

Name	*Year Founded*	*Silicon Valley Members*	*Mission*
Chinese Institute of Engineers (CIE/USA)	1977	1,000	A technical organization that promotes communication and interchange of information among Chinese engineers and scientists. *www.cie-sf.org*
Asia America MultiTechnology Association (AAMA)	1979	1,100	A multi-technology business network promoting the success of the region's Asian-American technology enterprises. *www.aamasv.com*
Silicon Valley Chinese American Computer Association (SVCACA)	1980	500	Sharing information, pooling resources, and cooperating with one another in order to leverage collective buying volume when dealing with vendors as well as with service companies. *www.svcaca.com*

Name	*Year Founded*	*Silicon Valley Members*	*Mission*
The Photonics Society of Chinese-Americans, Northern California Chapter	1985	N/A	Provides a monthly seminar that covers various topics that deal with light; provides a way to meet and network with other professionals in the Bay Area. *www.eoa-psc.org/*
Silicon Valley Indian Professionals Association (SIPA)	1987	2,300	A forum for young entrepreneurial expatriate Indians to contribute to cooperation between the United States and India in technology areas. *www.sipa.org*
Chinese Software Professionals Association (CSPA)	1988	2,000	Promotes technology collaboration and facilitates information exchange in the software profession. *www.cspa.com*
Silicon Valley Chinese Engineers (SCEA)	1989	4,000+	Promotes entrepreneurship and professionalism among members and seeks to establish ties to mainland China. *www.scea.org*
Monte Jade Science and Technology Association (MJSTA)	1990	470 individual 180 corporate	Promotes business cooperation and long-term investment between Taiwan and the San Francisco Bay Area. *www.montejade.org*
Chinese American Semiconductor Professionals Associations (CASPA)	1991	3,500	Promotes technical knowledge, communication, information exchange, and collaboration among semiconductor professionals. *www.caspa.com*

Name	*Year Founded*	*Silicon Valley Members*	*Mission*
North America Taiwanese Engineers Association (NATEA)	1991	600	Promotes technological knowledge and its applications and management training opportunities for Taiwanese Americans. *http://natea.org*
The Indus Entrepreneur (TiE)	1992	2,000	Fosters entrepreneurship by providing mentorship and resources. *www.tie.org*
Chinese Information and Networking Association (CINA)	1992	2,600	Chinese professionals who advocate technologies and business opportunities in information industries. *www.cina.org*
North American Chinese Semiconductor Association (NACSA)	1996	1,000+	Promotes professional advancement in the semiconductor sector; interaction between the United States and mainland China. *www.nacsa.com*
Silicon Vikings	1997	1,575	Provides networking opportunities for people with an interest in Sweden, technology, and business and finance. *www.siliconvikings.com/*
Sivan Group (formerly known as CarmelDalia)	1997	N/A	A not-for-profit forum of leading Israeli and Israel-affiliated business executives in Silicon Valley that provides networking opportunities to encourage and facilitate helping the Israeli high-tech community. *www.sivangroup.org/*

Name	*Year Founded*	*Silicon Valley Members*	*Mission*
Chinese Bioscience Association (CBA)	1998	N/A	Aims to enhance information exchange in the biotechnology, biopharmaceutical, and related businesses such as future technology trends, entrepreneurship, and the like. *www.cbasf.com/*
Hua Yuan Science and Technology Association (started as "E-Club" or Chinese Entrepreneurs Club)	1999	2,500	Promotes the technological, professional, and scientific development of the Chinese business community and the cross-border business relationship between Silicon Valley and mainland China. *www.huayuan.org*
Asia-Silicon Valley Connection (ASVC)	2000	N/A	Promotes the linkage of entrepreneurial initiatives between Asia and Silicon Valley. *www.asvc.org/*
Korean American Society of Entrepreneurs (KASE)	2000	N/A	A network of Korean Americans interested in starting or playing key roles in professionally managed high-growth companies in the United States. *www.kase.org*
Silicon Valley Chinese Wireless Technology Association (SVC Wireless)	2000	N/A	A wireless-focused organization designed to appeal to anyone interested in cutting-edge wireless technologies and market developments. *www.svcwireless.org*

Name	*Year Founded*	*Silicon Valley Members*	*Mission*
Armenian High Tech Council (ArmenTech)	2000	N/A	Aims to promote and support the creation and development of technology-based businesses in Armenia. *www.armentech.org*
Vietnamese Silicon Valley Network (VSVN)	2001	N/A	Provides the opportunity for entrepreneurs, investors, and professionals to interact, exchange ideas, and share feedback. *www.vsvn.org*
Hispanic Network of Entrepreneurs (Hispanic-Net)	2001	N/A	Supports Hispanic entrepreneurs, executives, and professionals in high-tech companies to advance business and personal opportunities. *www.hispanic-net.org*
Korean Information Technology Network (KIN)	2001	N/A	Promotes exchange of ideas and opportunities and access to resources to achieve preeminence in information technology. *www.koreanit.org*
American Business Association of Russian Expatriates (AmBAR)	2002	N/A	Aims to create an informal networking platform for the professional community of Russian expatriates in the United States; and to enhance cooperation between the United States and Russia in high-tech areas. *www.ambarclub.org/*
Silicon Iran	2002	7,703	A technical magazine serving the high-tech Iranian community globally. *www.siliconiran.com*

Name	*Year Founded*	*Silicon Valley Members*	*Mission*
Silicon Valley Japanese Entrepreneur Network (SVJEN)	N/A	N/A	A network to support start-up activities of professionals inspiring to become entrepreneurs by exchanging business information. *www.svjen.org/*
ANZA Technology Network	N/A	N/A	Aims to build the most extensive network of Australian, New Zealand, and U.S. companies and executives in the technology sector. *www.anzatechnet.com/*
SiliconFrench	N/A	N/A	Helps SV French-speaking professionals' personal and professional development through exchange of knowledge, information, and contacts. *www.siliconfrench.com/*
Chinese American Information Storage Society	N/A	N/A	Aims to increase knowledge in technology, applications, and business opportunities in the information storage industry through education, including seminars and conferences. *www.caiss.org/*

APPENDIX B

Survey Results: Immigrant Professionals in Silicon Valley

The results of a 2001 survey of members of seventeen Silicon Valley immigrant professional associations (Saxenian 2002) provide strong evidence of the growth of brain circulation and the emergence of cross-regional business and technical networks linking Silicon Valley and other regions, particularly in Asia. Approximately 80 percent of the foreign-born engineers and professionals in Silicon Valley who responded to the survey (21 percent of the total invited to participate) reported that they exchange information about jobs and business opportunities in the United States as well as about technology with colleagues in their home countries. And more than 20 percent of the Indian and Chinese respondents reported exchanging such information on a regular basis.

More than 40 percent of all foreign-born respondents reported that they travel to their home country for business at least once a year, and approximately 5 percent reported making the trip five or more times per year. Many work for businesses with activities in both regions. Others participate in conferences, technical seminars, or research and teaching opportunities sponsored by government agencies, universities, or professional associations. Taiwanese respondents stood out in this regard, with only 36 percent reporting that they *never* traveled home for business, compared to 56 percent of Chinese and 48 percent of Indian respondents who never traveled home.

Three-quarters of the foreign-born professionals surveyed report that they know at least one person who has returned permanently from the United States to their home country—and a small percentage (China, 6 percent; India, 4 percent) report knowing ten or more. Taiwan is a clear exception: 16 percent of respondents know ten or more returnees, and only 13 percent don't know any. This reflects the growing economic opportunities in Taiwan and the related professional and business ties that have developed between Silicon Valley and Taiwan. It is likely that more recent survey data would show greater traffic between Silicon Valley and India and China as well as higher return rates; anecdotal evidence suggests that growing numbers of young Indian and Chinese professionals returned home as the technology downturn worsened in 2002 and 2003. Accurate data, however, are not yet available.

Even those foreign-born engineers and executives who remain in Silicon Valley full time are often well integrated into the transnational business community. A sizable proportion of the survey respondents from India (48 percent), Taiwan (42 percent), and China (34 percent) reported that they had helped to arrange contracts with businesses in their home countries. Some also moonlight as consultants for firms in their native countries, providing product development or marketing advice, or helping to arrange business contracts. Smaller but still significant proportions of respondents had advised businesses in their home countries (India, 34 percent; Taiwan, 24 percent; China, 15 percent) or invested their own money in businesses in their native country (India, 23 percent; Taiwan, 17 percent; China, 10 percent). At least one-quarter of the foreign-born respondents reported meeting with government officials from their home countries, although less than 5 percent reported regular meetings with these officials. More than 40 percent of the Indian and Chinese respondents indicated that they would consider returning to live and work in their country of birth if the professional opportunities were available; and these proportions were, not surprisingly, higher for those under age 35.

The survey shows that involvement in local networking and entrepreneurship as well as transnational activities—from cross-regional travel to advising or investing in businesses to meeting with top government officials—is closely correlated with the age of the individual. It appears that there is a core group of more experienced and older Indian and Chinese immigrants (age 50 or over) in Silicon Valley who are actively involved in participating in local associations, starting businesses, and building connections to their home countries—although this age group is also the least likely to return home permanently.

The survey results suggest that the rate of entrepreneurship among foreign-born engineers is comparable to that of the native-born population. And although a majority of Indian and Chinese entrepreneurs go into business with one or more colleagues from their home country, this ethnic dominance decreases as the company grows. In most other aspects, immigrant technology businesses in Silicon Valley appear to be similar to the mainstream companies: both foreign- and native-born entrepreneurs report raising comparable amounts of capital, initially from angel investors and later from venture capital firms. And both groups report that the most significant challenge in raising capital is access to venture capitalists.

In sharp contrast to their native-born counterparts in Silicon Valley, however, half of the foreign-born entrepreneurs with businesses in the Valley had also set up subsidiaries, joint ventures, subcontracting, or other business operations in their home countries to exploit their privileged access to the market, low-cost skill, and other resources. A majority of the remainder say that they would consider establishing such operations in the future. These operations are concentrated in a small number of fast-growing urban areas, and their specialties reflect those of the economies in which they are located. In the Greater China region, for example, these businesses are primarily involved in marketing, sales, and hardware design and manufacturing, while in India the focus is primarily on software or content development and software services.

The survey also included a set of questions about the factors shaping decisions to return home to work. More than three-quarters (78 percent) of the respondents from Greater China ranked "professional opportunities" as a very important factor, significantly ahead of the next-ranked factor ("culture and lifestyle," with 62 percent of respondents considering it very important). By contrast, 75 percent of Indian respondents ranked "culture and lifestyle" as very important to the decision to return, followed closely by "professional opportunities" (68 percent). 60 percent of Indian respondents also ranked "opportunity to contribute to home country" as very important, compared to 46 percent of those from Greater China. "Favorable government treatment" was considered important by less than half of the Chinese (47 percent) and Indian (49 percent) respondents, suggesting that government incentives are not the deciding factor for most individuals who are considering returning to their home countries to work. Finally, "barriers to professional advancement," or the glass ceiling, were ranked as very important by 44 percent of those from Greater China and only 26 percent of Indians.

A set of questions was addressed to the subset of respondents who were on the founding team of a start-up (n = 452). The first questions asked those with business operations in their home country what factors influenced the decision to locate there. The factors ranked very important by the most Indian respondents were "availability of skill" (85 percent) and "low-cost labor" (73 percent), supporting the view that in the current economy the scarce resource is not simply cheap labor, but lower-cost skill. Three-quarters of all respondents from Greater China ranked "access to market" as very important in their decision to locate business operations in that region, compared to 18 percent of Indian respondents. The second and third most frequent factors ranked as very important by Chinese respondents were, as with the Indians, "low-cost labor" (46 percent) and "availability of skill" (36 percent). "Access to capital" was ranked as important by 32 percent of the respondents from Greater China, compared to only 6 percent of respondents from India. "Access to technology" was considered important by 27 percent of Indian and 18 percent of Chinese respondents. Consistent with the responses to earlier questions, only 16 percent of Chinese and 6 percent of Indian respondents considered "financial incentives from the government" an important factor in the decision to locate business operations in their home country.

Finally, respondents were asked what would deter them from starting a business in their home country. Not surprisingly, "unreliable infrastructure" was ranked by 74 percent of Indians as a very important deterrent, compared to only 22 percent of Chinese. Large proportions of both Indian (73 percent) and Chinese (58 percent) respondents ranked "government bureaucracy/regulations" as very important deterrents to starting a business in their home country. An "inadequate legal system" was also an important deterrent for 48 percent of Chinese and 32 percent of Indian respondents. "Political or economic uncertainty" was ranked as very important by 46 percent of Chinese, but only 24 percent of Indian respondents; and an "immature market" was considered important by 34 percent of Chinese and 27 percent of Indians. Similarly, 32 percent of Chinese respondents ranked "unfair competition" as a very important deterrent, compared to only 13 percent of Indians. By contrast, 29 percent of Indian respondents ranked "poor business services" as an important deterrent, compared to 15 percent of Chinese. The final two factors, "inferior quality of life" and "lack of access to capital," were considered very important to only approximately 20 percent of respondents from both groups. In sum, for Indians, the key obstacles to starting a business at home were the infrastructure and government bureaucracy/regulation, while for those from Greater China, the most important obstacles were government bureaucracy/regulation, an inadequate legal system, and political and economic uncertainty.

NOTES

INTRODUCTION

1. Women are better represented among foreign-born scientists and engineers in Silicon Valley than among their native-born counterparts, particularly for immigrants from China and Taiwan. In 2000, 21 percent of the total S&E workforce in SV was female (with 19 percent of the native-born and 23 percent of the foreign-born workforce). Women accounted for almost one-third (32 percent) of the China-born and 30 percent of the Taiwan-born S&E workforce. By contrast, women accounted for only 15 percent of the region's S&E workers born in India (U.S. Census Bureau 2003).

1. SURPRISING SUCCESS

1. Hobday 1995.

2. Bordo et al. 2003; Zeitlin and Herrigel 2000; Sowell 1996; Brenner 1994.

3. Todaro 1985, p. 123.

4. Rapaport 2002.

5. Bhagwati 1985.

6. Chandler 1977; see also Zuboff 1988; Blaug 1986; Piore and Sabel 1984.

7. Zeitlin and Herrigel 2000; Chandler and Daems 1980.

8. Gerschenkron 1962.

9. Waldinger et al. 1990; Light and Bonacich 1988.

10. Sabel and Zeitlin 1985; Galbraith 1967.

11. Herrigel 1996; Scranton 1983; Brusco 1982; Chinitz 1961.

12. Chen 2002; Cheng 2001; Cheng and Gereffi 1994.

13. Gold 1986; Barrett and Whyte 1982.

14. Amsden 2001, 1989.

15. Arthur 1994; Krugman 1991.

16. Porter 1998.

17. The industry cluster has become a standard category for both analysis and policymaking, in spite of its analytical limitations. For a review of the now voluminous literature, see Cortright (2005).

18. Piore and Sabel 1984.

19. Borrus et al. 2000; Bartlett and Ghoshal 1998; Dedrick and Kraemer 1998; Gereffi and Korzeniewicz 1994.

20. Saxenian 1994.

21. Ferguson 1990; Borrus 1988.

22. Gompers et al. 2003.

23. Quotations from *www.svb.com.*

24. Interview with Min Zhu, April 16, 2001.

25. Moore 1996, p. 167.

26. Young 1928.

27. Zhang 2003.

28. Joint Venture: Silicon Valley Network 2004.

29. Leslie and Kargon 1996, p. 470; see also Lenoir et al. 2004.

30. Fallick et al. 2003.

31. Gilson 1998.

32. Interview with John Shoch, August 29, 1998.

33. David Wang, executive vice president of the semiconductor equipment manufacturer Applied Materials, provided the following list of promising IC niche market opportunities to aspiring entrepreneurs at a conference in 2003: LCD Display Controller/Driver ICs; Multimedia ICs: MPEG video/audio decoder, TV decoder, camera controller, decoder, image processor, DVD/VCD controller/driver, MP3 decoder; Cellular Handset

Peripheral ICs: power management ICs, audio encoder/decoder ICs; and PC Peripheral ICs: mouse/keyboard controller, CD ROM/DVD ROM controller, core logic chipset, Ethernet/IEE1394 controller, scanner controller (NACSA 2002–2003 Annual Report, 30).

34. Cole 2003.

35. Ernst 1994.

36. Lee 2005.

37. Shih 1996, p. 91.

38. Kogut and Zander 1993.

39. Comments of Bill Ting at Hewlett-Packard SSAE conference, October 19, 1998; notes taken by the author.

40. Chesbrough 2003; Ashkenas et al. 2003.

2. LEARNING THE SILICON VALLEY SYSTEM

1. Hill 1996.

2. The U.S. Immigration Act of 1965 or Hart-Celler Act established the H-1B visas; Taiwan, for example, which was historically limited to a quota of one hundred immigrant visas per year, sent only forty-seven scientists and engineers to the United States in 1965. Two years later, the number had increased to more than thirteen hundred (Chang 1992).

3. Burton and Wang 1999.

4. China- and India-born immigrants accounted for 20 percent and 16 percent, respectively, of all foreign-born U.S. residents with science and engineering doctorates in 1999. Taiwan-born immigrants accounted for another 6 percent (NSB 2002).

5. In 1990, Chinese- and Hong Kong-born accounted for 3.2 percent of Silicon Valley's 131,585 managers and professionals, followed by Taiwanese at 3.0 percent, Indians 2.8 percent, Vietnamese 1.6 percent, and Philippine-born 1.6 percent. The only other national groups with more than 1 percent of the total were the U.K., Germany, Iran, and Japan. U.S. Census Bureau 1990. 1990 Census of Population and Housing, Public Use Microdata Samples (PUMS).

6. Interview with Lester Lee, July 1, 1997; interview with David Lee, January 22, 1997.

7. Asian Americans for Community Involvement 1993.

8. Saxenian 1999.

9. Fernandez 1998; Tang 1993.

10. Interview with Satish Gupta, May 29, 1997.

11. The density of each of these communities in the Silicon Valley area means that there are many other forums for socializing and building ethnic identities. For example, more than ninety Chinese schools in the Bay Area teach Mandarin to school-age children. These schools play an important role in the Chinese community, as do certain churches, table tennis tournaments, and so on. The Indian community meets frequently in local cricket clubs, temples, and in other religious and political groups as well as cultural events. Immigrants also select places to live based on their ethnic social networks, and the residential data confirms that both Indians and Chinese are highly concentrated in certain cities in the South Bay area.

12. Immigrant entrepreneurs in Silicon Valley are mainly male; however, I interviewed the following women: Pauline Lo Alker, President and CEO, Amplify.net; Priscilla M. Lu, Founder and Chairman of Board, interWAVE Communications Intl.; Vani Kola, Founder and CEO, Nth Orbit; Vinita Gupta, President and CEO, Quick Eagle Networks (formerly Digital Link Corporation); and Radha Basu, Chair and CEO, SupportSoft, Inc.

13. Granovetter 1995.

14. Interview with Victor Wang, conducted by Guoqiang Yu, July 24, 1999.

15. Interview with Jason Wu, April 16, 2001.

16. Interview with Fred Cheng, March 25, 1997.

17. Interview with Jimmy Lee, March 5, 1997.

18. Tsinghua University is China's most prestigious engineering school. It was founded in 1911 in Beijing as a preparatory school for students being sent by the government to study in the United States. The Taiwan branch was established after 1949, and the two branches have maintained academic ties since the 1950s. Note that because of different transliteration standards, the English spelling of the Taiwan branch is Tsinghua, while the spelling of the mainland branch is Qinghua.

19. Yoshida and Leopold 2001a, 2001b.

20. Interview with Jerry Shiao, March 11, 1999.

21. Naipaul 1992.

22. Interview with Ronald Chwang, March 25, 1997.

23. Interview with Jesse Chen, June 7, 1997.

24. Interview with Mohan Trika, June 3, 1997.

25. Interview with Albert Y. C. Yu, March 5, 1997.

26. Interview with David Lam, January 24, 1997.

27. Interview with Gerry Liu, January 22, 1997.

28. Interview with Chandra Shekar, April 8, 1997.

29. Interview with Vinod Khosla, January 14, 1997.

30. Interview with Satish Gupta, May 29, 1997.

31. Interview with Shelby Chen, May 24, 2001.

32. Committee of 100 White Paper, 2001.

33. Saxenian 2002.

34. See Saxenian 2002, Figure 5.4.

35. Light and Roach 1996.

36. The Dunn & Bradstreet surname search may still understate immigrant entrepreneurship in the region because firms started by immigrants with non-Asian CEOs are not counted. Interviews suggest that this was frequently the case in the 1970s and 1980s in Silicon Valley, where venture capital financing was often tied to the requirement that non-Asian senior executives be hired. This seems a more likely source of bias than the opposite scenario, that is, firms started by non-Asians being required to hire a Chinese or Indian CEO.

37. Saxenian 2002.

38. Ibid.

3. CREATING CROSS-REGIONAL COMMUNITIES

1. Silicon Valley Bank Web site: *www.svb.com/services/intlnetworks.asp;* access date: August 20, 2004.

2. More than 20,000 highly skilled Irish workers moved to the United States between 1980 and 1995—a sizable proportion of the most highly educated youth in a country of 3.5 million people. In the mid-1990s the flow

reversed, and the Dublin urban area was transformed by the growth of technology entrepreneurship. By the end of the decade Ireland was the second-largest exporter of packaged software in the world, with local software firms employing more than 10,000 workers and receiving proportionately the highest level of venture capital investment in the European technology sector (O'Riain 2002; International Venture Capital Association 2002). As a result, the All Ireland–U.S. Chamber of Commerce Web site reported in 2003: "Every year there is more interaction between businesses in Ireland and the United States. Starting with Silicon Valley, the synergy between industry and Ireland is unmistakable" (*www.irishlinx.com;* access date: August 27, 2004).

3. National Science Board 2004, Appendix Table 2-31.

4. National Science Board 2002. Chinese universities also granted more than 6,000 S&E doctorates annually in the late 1990s. Indian universities granted more than 4,000 doctorates, but they were almost entirely in the natural sciences and agriculture rather than engineering. "Science & Engineering Indicators—2002," *www.nsf.gov/sbe/seind02;* access date: May 2, 2002. First university degrees in S&E: Appendix Table 2-18; doctoral S&E degrees: Appendix Table 2-42.

5. Goering 2004.

6. U.S. Bureau of Economic Analysis, n.d.

7. Chwang 1997.

8. TiE now has branches all over Europe and Asia, and its current mission statement makes no reference to South Asia, reading: "TiE is a global, not for profit network of entrepreneurs and professionals dedicated to the advancement of entrepreneurship"; and its Web site states that TiE stands for "Talent, Ideas, and Enterprise." However, TiE's leadership—directors, board of trustees, and charter members—are all of Indian (or South Asian) origin, and meetings are attended overwhelmingly by members of the Indus community.

9. Chugani 2002.

10. See *www.siliconindia.com.*

11. Krishnadas 2004, p. 70.

12. Ibid.

13. "Like Company Like Bank," 2000.

14. Ibid.

15. Interview with Daniel Quon, July 1, 1997.

16. Interview with Carmen Chang, April 26, 2001.

17. Hellman 1998, p. 2.

18. Interview with Herbert Chang, July 22, 1997.

19. The following Silicon Valley venture firms have invested actively in Asia since at least the 1990s: AsiaTech Ventures, Crimson Venture, Crystal Internet Ventures, H&Q Asia Pacific, IDG Ventures, and the WI Harper Group. A large number of mainstream venture capital firms started investing in China after 2001, resulting in a boom market that is reminiscent of the late 1990s in the United States. In 2002 there were only 38 venture firms operating in China; a year later there were 172. Mark Marshall, "Venture capitalists across Silicon Valley are falling for China," *The San Jose Mercury News,* October 10, 2004. While the trend is newer, several Silicon Valley venture capital firms now focus on India, including Jumpstartup, New Path Ventures, and Westbridge Capital.

20. Interview with C. B. Liaw, August 28, 1996.

21. The percentages are as follows: from 1999 through 2002 Israel's venture capital investment as a percentage of GDP was 0.6 percent, while the figures for the United States and Canada were 0.4 and 0.33 percent respectively (OECD Venture Capital database, cited in Global Insight 2004).

22. *Asia Venture Capital Journal* and *Financial Express* (China Business Review, July-August 2004).

23. *The Economist* 2004, p. 98. Although Taiwan led Asian private equity investment in the 1990s, these investments have shifted across the Straits in recent years. In 2003 China and Hong Kong led the rest of Asia in private equity fundraising with 35 percent of the total, followed by South Korea (24 percent), India (12 percent), and Taiwan (11 percent). The nations in the region were Australia, China, India, Indonesia, New Zealand, the Philippines, Malaysia, Singapore, South Korea, Taiwan, Thailand, and Vietnam (Thomson Venture Economics 2004).

24. Interview with K. Y. Han, May 13, 1997.

25. Tony Sitathan, "The Five Hot Shots: Right Place, Right Time," *Electronic Business Asia,* December 1998. *www.eb-asia.com;* access date: January 26, 2000.

26. IMD International 2004; Trajtenberg 2001.

27. Trajtenberg 2001; Breznitz 2000; Teubal et al. 2000.

28. Cohen and Haberfeld 2001; Paltiel 2001.

29. deFontenay and Carmel 2004; Mahroum 1998.

30. Autler 2000, p. 49.

31. Tania Hershman, "Tech's New Promised Land," *Wired,* January 17, 2000.

32. Rock 2001; Autler 2000.

33. Autler 2000.

34. Interview with Fred Adler, February 9, 1999.

35. Rock 2001.

36. Ibid.

37. Khavul 2003.

38. IVC Research Center 2004.

39. Interview with Itzhak Levit, May 12, 2003.

40. Stub 2003, p. 16.

41. Blackburn 2003, p. 12.

42. Autler 2000, p. 43.

43. Paltiel 2001, Figure 6.

44. Angwin 1997.

45. Paltiel 2001, p. 14; Angwin 1997; Zoellner 2000, p. B6.

46. Autler 2000; Gillmor 1998.

47. Interview with Itzhak Levit, May 12, 2003.

48. Breznitz 2000.

49. Stub 2003, p. 16.

50. Autler 2002, p. 11.

51. Khavul 2003, p. 24.

52. Trajtenberg 2000.

53. Singapore has attracted substantially more foreign direct investment

per capita than any other major Asian economy. All the data in this paragraph are taken from Ben Dolven, "Taiwan's Trump," *Far Eastern Economic Review,* August 6, 1998, pp. 12–15.

4. TAIWAN AS SILICON SIBLING

1. Taipei 2000.

2. Taiwan National Science Council 1995.

3. Trajtenberg 2000. Taiwan's Ministry of Economic Affairs reports that in 2003 Taiwan ranked fourth in the world in terms of the total number of invention patents registered in the United States, third in the world in terms of the number of patents registered per million persons, and fourth in the world in terms of patent intensity. These patents were especially concentrated in the fields of semiconductor production process equipment and LCD display technology. STICNET, Science and Technology Information Center, National Science Council, Taiwan, August 19, 2004.

4. Wade 1990.

5. Mathews 1997.

6. Amsden and Chu 2003.

7. Ernst 2000; Dedrick and Kraemer 1998.

8. Quotation from Callon 1995.

9. Companies like First International Computer are not generally included in these figures because they produce boards that they sell within completed systems. If producers that sell systems (from bare-bones to finished goods) with motherboards were included, Taiwan's share of the market would be well over 80 percent.

10. Levy and Kuo 1991.

11. Chung 2000.

12. Phred Dvorak and Evan Ramstad, "Flat-Panel Consolidation," *Wall Street Journal,* February 8, 2005.

13. Carrie Yu, "Interview with TPV Technology, World's Number One LCD Monitor Supplier," *DigiTimes,* September 10, 2004. *www.DigiTimes.com;* access date: November 15, 2004.

14. In 2003 Acer surpassed Sony to become the world's sixth-largest notebook vendor, and Japan's market share in notebook computers fell from 36 percent in 2002 to 26 percent (Tzeng 2003).

15. Hobday 2001; Best 2001; McKendrick et al. 2000.

16. Rasiah 2002, 2001.

17. Callon 1995.

18. Ibid., p. 8.

19. Wade 1990, p. 88.

20. Wade 1990.

21. Interview with Patrick Wang, May 21, 1997.

22. Interview with Genda Hu, May 13, 1997.

23. This section draws heavily from the account provided in Meany (1994). For more detailed accounts of the development strategies for Taiwan's technology industry, see Liu 1993; Chang, Shih, and Hsu 1994; and Chang, Hsu, and Tsai 1999.

24. Y. S. Sun was trained as an engineer in China and studied at the Tennessee Valley Authority in the United States in the 1940s. After working for many years at Taiwan Power Company, he served as Taiwan's Minister of Communications in 1966–1969, as Minister of Economic Affairs in 1969–1978, and as Premier in 1978–1984.

25. METS continues to meet every other year in Taipei. The session topics for the twentieth meeting in 2004 included nano- and MEMS technologies, broadband communications and network systems integration, resource reuse and recycling, management of knowledge and digital technology, industrial transformation and transition, flat-panel displays and optoelectronics, and biotechnology.

26. Chang and Hsu 2001, p. 290.

27. Tzeng and Lee 2001, p. 311.

28. Ibid., p. 319.

29. Interview with Ken Tai, May 16, 1997.

30. Wade 1990.

31. Shih 1996, pp. 175–176.

32. Hsu 1997.

33. For example, Yang Ding-yuan became president of Winbond Elec-

tronics; Robert Tsao became president of UMC; Shih Chin-tay rose to senior levels in ERSO and eventually became president of ITRI; and C. C. Chang became director general of ERSO.

34. Mathews 1997.

35. Interview with Ta-lin Hsu, June 11, 1997.

36. Linden 2000.

37. Chinese Institute of Engineers USA 1999.

38. Addison 2001, p. 55.

39. In U.S. dollars, NTD (New Taiwan Dollars) $800 million is approximately $24 million.

40. Interview with Ta-lin Hsu, June 1, 1997.

41. Howell et al. 2003, p. 35.

42. The founders or presidents of some of Taiwan's leading corporations, including TSMC (Morris Chang), Mosel (Nasa Tsai), Etron (Nicky Lu), Silicon Integrated Systems (S. Samuel Liu), Chung-Chuang Display (Hsing-Chien Tuan), and Micronix (Miin Wu and Ming-Chu Wu) all hold graduate degrees from Stanford, mainly doctorates. The founder of Mitac (Matthew Miau), chairman of the Grace Group (Winston Wang), chairman of First International Computer (Ming J. Chien), and managing director of Walden International in Taipei (Tzu-Hwa Hsu) are U.C. Berkeley graduates, and Patrick Wang, founder of Microelectronics Technology, attended both Stanford and U.C. Berkeley. Winbond founder Ding-Yuan Yang and ITRI director Chintay Shih are both Princeton graduates. The president of TSMC (Rick Tsai) and the chairman of CSMC (Peter Chen) both hold doctorates from Cornell University. The president and CEO of VIA (Ming J. Chien), chairman of SMIC (Richard Chang), president of Altek (Alex Hsian), and vice president of CSMC (Wilson Yu) all hold graduate degrees from other U.S. universities.

43. Interview with Genda Hu, May 13, 1997.

44. Interview with Robert Tsao, May 19, 1997.

45. Interview with Morris Chang, May 12, 1997.

46. Science Park Administration 2001.

47. Tanzer 1998.

48. Luo and Wang 2001.

49. Tanzer 1998.

50. Ibid.

51. National Youth Commission 2000.

52. National Science Board 2002.

53. Mathews 2002.

54. Shih 1996, p. 94.

55. Ibid., pp. 19–20.

56. Shih 1996.

57. Lin 2000, p. 203.

58. Ibid., p. 220.

59. Shih 1996, p. 197.

60. Multitech and Mitac were cloning Apple II at the time, and Compeq was still making calculators and had no experience with circuit boards.

61. Chang et al. 1999.

62. Kishimoto 2003.

63. Lin 2000.

64. Tzeng and Lee 2001, p. 334.

65. *www.tomshardware.com.* Other prominent Ace spin-offs include Ben Q and AUOptronics.

66. Chen and Wang 2001, p. 77.

67. Shih 1996, p. 11.

68. Ibid., p. 16.

69. Ernst 2000.

70. Interview with Eric Wang, May 15, 1997.

71. *Common Wealth* 1996.

5. TAIWAN AS PARTNER AND PARENT

1. Interview with C. S. Ho, May 15, 1997.

2. Interview with Ken Tai, May 16, 1997.

3. Interview with Eric Wang, May 15, 1997.

4. In 1999 Accton Technology, another Taiwan-based network product manufacturer, invested in three start-ups in Silicon Valley (Altima, Narvell, and Rapidstream) to obtain transceiver, firewall, digital signal pro-

cessor, and virtual personal network technologies. This explains why Accton's R&D vice president flies to Silicon Valley as frequently as twice a month. Like D-Link, Accton has invested in local companies including ADM Tek for IC design and OLE for cable modem technology. Interview with Jinn Yuh, August 19, 1999.

5. "D-Link Corporation announces [transfer of] Alpha Networks' shares to Quanta Computer," March 11, 2004, D-Link press release.

6. Cameo Communications, one of D-Link's earliest partners, remains one of its lead suppliers. Other suppliers include BroadWeb, Cipherium Systems, and L-7 Networks, which provide network-security solutions; DrayTek, which specializes in voice-over IP (VOIP) technology; Z-Com, producer of WLAN modules such as 802.11 WLAN cards; and Atheros, a designer of 802.11 WLLAN chipsets.

7. As early as 1994, the average hourly wages of both unskilled production workers and engineers in Taiwan had more than doubled, to 32 percent of U.S. wages from 15 percent in 1985. Wages of Taiwanese production workers were on par with those in Korea and Singapore and five times those in Malaysia and Thailand (Chang et al. 1999).

8. Interview with Ken Tai, May 16, 1997.

9. "The Flexible Tiger," 1998, p. 73.

10. BSMI 2000.

11. Kishimoto 2003.

12. Shelley Neumeier, "Going to China Has Become a Competitive Imperative," *Corporate Board Member,* May/June 2004, pp. 24–29.

13. Interview with George Huang, May 22, 1997.

14. Hall 2000.

15. Ehrgood 2000.

16. David Tzeng, "HP praises Taiwan Partners," *DigiTimes,* October 19, 2004. *www.digitimes.com;* access date: November 15, 2004.

17. Hagel 2004.

18. Interview with K. Y. Han, May 13, 1997.

19. "Spirit of Partnership: Turning Industry Cliché into a Successful Business Model." *www.xilinx.com/company/about/partner;* access date: April 23, 2005.

20. Ibid.

21. Interview with Steve Tso, March 15, 1999.

22. Interview, Dr. David Wang, "Global Chipmakers Gather in China," *Nikkei Electronics Asia,* March 2001. *www.nikkeibp.asiabiztech.com/nea;* access date: March 28, 2001.

23. ASE Inc., 2002.

24. Minamikawa 2000.

25. Stanford Alumni Conference 2000.

26. Worldwide foundry revenues were over $22 billion in 2004, with pure-play firms accounting for more than $18 billion and Integrated Device Manufacturers (IDMs) accounting for the balance. Foundry services are provided by both pure-play firms that dedicate their capacity to providing foundry services and IDMs that provide foundry services as a way to extend the useful life of fabs with trailing-edge process technologies.

27. Hagel 2004.

28. Luo and Wang 2003.

29. Hall 1999.

30. R&D investments are higher in the IC manufacturing and design sector than in the more cost-driven PC component industries. *www.sipa.gov.tw/en/;* access date: November 15, 2004.

31. Hagel 2004.

32. Hamilton 1997.

33. Tong and Wang 2000.

34. Guan 2002; Landler 2001; Flannery 1999.

35. Adam Hwang, "Taiwan: China-based manufacturing accounted for 85.1 percent of 2004 IT hardware production value," *DigiTimes,* March 3, 2005. *www.digitimes.com;* access date: March 4, 2005.

36. Ernst and Kim 2002.

37. "Motherboard Industry Overview: Production and Capacity," *DigiTimes,* January 22, 2005. *www.digitimes.com;* access date: March 25, 2005.

38. Skanderup 2004.

39. This industrial policy dialogue has been described by Dieter Ernst (2003) as "multi-level strategic decision-making" that operates on four

levels, from building a consensus on a long-term vision to the implementation and evaluation of specific projects. This institutional innovation, in Ernst's view, allows for "a high degree of resourcefulness and versatility in mobilizing multiple upgrading actors and their resources, as well as in soliciting policy contributions from outside of the state apparatus and from outside Taiwan's borders" (p. 29). As a small and politically isolated economy, Taiwan relies particularly heavily on these international connections.

40. Lin 2000; Hsu 1997.

41. OECD 2002.

42. Ernst 2003.

43. Mike Clendenin, "Creative play all in a day's work at ITRI," *Electronic Engineering Times,* August 16, 2004, pp. 18–24.

44. Clendenin 2004.

45. Chris Hall, "Flat panels go nanotech: Taiwan's ITRI pioneers CNT-FED, part three," *DigiTimes,* January 25, 2005. *www.digitimes.com;* access date: January 25, 2005.

46. Daniel Shen and Jessie Shen, "Broadcom talks about its Taiwan R&D center," *DigiTimes,* November 16, 2004. *www.digitimes.com;* access date: November 16, 2004.

47. Bruce Einhorn and Peter Burrows, "Taiwan's Climb up the Tech Ladder," *Business Week,* October 6, 2003.

48. CENS, "Southern science park projects record revenue," December 14, 2004. *http://news.cens.com;* access date: December 16, 2004. See also *www.sipa.gov.tw* for data on both parks.

49. Michael McManus, "The XPC factor: An interview with Ken Huang, the man behind the Shuttle XPC," *DigiTimes,* February 1, 2005. *www.digitimes.com;* access date: February 7, 2005.

50. Vyacheslav Sobolev, "Microdriving the future—An interview with Cornice VP David Feller," *DigiTimes,* March 11, 2005. *www.digitimes.com;* access date: March 11, 2005.

51. "Foundry industry overview," *DigiTimes,* January 2005, *www.digitimes.com;* access date: July 20, 2005.

52. Adam Hwang, "MIC: Taiwan continues to dominate global IT and networking hardware supply," *DigiTimes,* March 4, 2005. *www.digitimes.com;* access date: July 20, 2005.

6. MANUFACTURING IN MAINLAND CHINA

1. "Overseas Chinese Head Home," 2001.

2. Data from the Ministry of Commerce show that foreign-owned companies accounted for 84.6 percent of high-tech exports from China in 2003. "Technology in China: The Allure of Low Technology," *The Economist,* December 20, 2003.

3. China Statistics Press 2003. The Beijing region includes Tianjin and Hebei provinces; the Shanghai region includes Shanghai, Jiangsu, and Zhejiang provinces.

4. Naughton and Segal 2001.

5. National Science Board 2004.

6. U.S. Embassy, Beijing, 2002.

7. Kahn 2004.

8. Ibid.

9. The American National Standards Institute, the leading U.S. standards group, issued a white paper in 2004 on behalf of the IT, electronics, integrated circuit, and telecommunications industries, criticizing China's closed standard-setting process.

10. Chen 2004.

11. Guochu and Wenjun 2001. These authors estimate that China permanently lost 200,000 science and technology workers.

12. Jiang 2001.

13. Howell et al. 2003, p. 113.

14. In a 2001 survey of immigrant professionals in Silicon Valley, 43 percent of respondents from mainland China reported that they would consider returning to live in China in the future. When asked to identify the factors that would influence their decision to return, the great majority (78 percent) ranked "professional opportunities" as very important. A majority (62 percent) also considered "culture and lifestyle" very important. How-

ever, other considerations appeared less important, including favorable government treatment (only 47 percent ranked this as important), making a contribution to their home country (46 percent), or barriers to professional advancement in the United States (44 percent). These factors were ranked from 8 to 10 on a ten-point scale, with 10 as the most important (Saxenian 2002).

The survey also sought to determine the motivations for return entrepreneurship. When asked what factors would figure most in the decision to start a business at home, more than 50 percent of the mainland Chinese respondents listed access to the market, the availability of skilled workers, and access to capital as the key factors. Conversely, when asked about problem areas that would deter them from starting a business in China, the factors most frequently mentioned were government bureaucracy and regulation, an inadequate legal system, and political or economic uncertainty.

15. Cooke 2004.

16. Murphy and Pao 2001.

17. Hamilton 2005.

18. These clusters of specialized producers are not unique to high technology. China's coastal cities are home to vast garment districts specializing in everything from socks and sweaters to neckties and underwear. David Barboza, "In Roaring China, Sweaters are West of Socks City," *New York Times,* December 24, 2004. Nor are they new. Debin Ma (2003) argues that Shanghai was the geographic center of industrialization in China during the first three decades of the twentieth century, with concentrations of modern industry and services that far surpassed the major British industrial cities such as Manchester and Glasgow. Data on officially approved Taiwanese investments in China are available online at ***http://www.chinabiz.org.tw/maz/*** *Eco-Month/1999, ~2000, ~2001,* and *~2002,* a Chinese-language site.

19. Jiang et al. 2004.

20. Zhejiang and its capital city, Hangzhou, are one of the wealthiest regions in China (after only Beijing and Shanghai) and are home to its largest concentration of privately-owned enterprises (91 percent of its 240,000 enterprises were privately owned in 2002). This ancient silk and textile region avoided domination by state-owned enterprises in the twentieth century be-

cause its proximity to the island of Taiwan led the Communist Party to avoid locating large state projects in the area. "A Survey of Business in China," *The Economist,* March 20, 2004, pp. 15–16.

21. National Jiao Tong University is the mainland counterpart of Taiwan's National Chiao Tung University. The name is the same in Chinese, but the English transliteration follows local preferences for its rendering.

22. "Letter from Taiwan," 2002.

23. Clendenin 2003a.

24. While costs in China vary depending upon the location and type of investment, the rule of thumb is normally as follows: land, owned by the state, is provided free for a negotiated term of lease; labor is 10 to 15 percent of the costs in Taiwan; and factory or facility construction is 20 to 25 percent of the costs in Taiwan (Cooke 2004, p. 500).

25. David Tzeng and Steve Shen, "Taiwan notebook makers may migrate all production to China," *DigiTimes,* November 2, 2004. *www.digitimes.com;* access date: November 2, 2004.

26. Cooke 2004.

27. Tzeng 2001.

28. Chwang 2003.

29. Naughton 2003, 1997.

30. Lin et al. 1998.

31. Naughton 2003.

32. For a detailed analysis of State Council Circular 18, which makes a very strong case that the VAT violates the WTO's national treatment requirement, see Howell et al. (2003).

33. Woetzel and Chen 2002.

34. Jiang Mianheng, who is quite active in the technology sector in China, has been dubbed "prince of Information Technology" because of his large investments in telecommunications, IC design and manufacturing, and several venture capital firms that in turn invest in technology ventures. He earned a doctorate in electrical engineering from Drexel University in 1991 and is vice director of the Chinese Academy of Science, but he is extremely secretive.

35. Howell et al. 2003, p. 93.

36. Hwa-Nien Yu, an SMIC adviser, has thirty-five years of IC R&D experience, including a position as a senior manager at IBM Research Center, and is currently chair of ITRI's Advanced R&D Steering Committee.

37. Naughton 2003. SMIC's recruitment of top managerial talent from TSMC has intensified the personal rivalry between Morris Chang, founder and CEO of TSMC, and Richard Chang, founder and CEO of SMIC. Both men returned to Taiwan after successful careers at Texas Instruments and started semiconductor companies. However, TSMC acquired Richard Chang's World Semiconductor Manufacturing Corp. and has recently filed a series of lawsuits against SMIC for theft of intellectual property.

38. Interview with Huann Min Tang, November 8, 2001.

39. "Searching for Hope," 2002.

40. Cai was previously deputy secretary of the Shanghai government, responsible for economic planning, finance, and research; he was awarded the title of "State-Class Economist"; and he is a member of the National Committee of the Chinese People's Political Consultative Conference. Cai was also deputy director of the Shanghai Planning Committee and Pudong Development Office. Discussion of the SMIC board can be found at *www.smics.com/website/enVersion/Investor/boarddirector.htm.*

41. The SMIC technology partnership with Singapore-based Chartered Semiconductor appears to be typical: Chartered transferred its 0.18-micron baseline logic process technology and patent rights to SMIC in exchange for a small equity stake and access to SMIC capacity for five years. The partnerships with Toshiba and Fujitsu are similarly structured (Howell et al. 2003, p. 96).

42. Maria A. Lester, "2003 Top Fabs." *Semiconductor International,* May 1, 2003. Available online: *www.reed-electronics.com/semiconductor/* Access date: July 1, 2004.

43. Data from PricewaterhouseCoopers 2004 and "SEMICON China" 2004.

44. PricewaterhouseCoopers 2004, p. 17, and Lee Han Shih, "Chinese Business: The Battle of the Fab Operators," *The Edge Singapore,* March 7, 2005.

45. Clendenin 2003c.

46. PricewaterhouseCoopers 2004.

47. Baijia 2003.

48. Howell et al. 2003.

49. "Shanghai Presents 156 Projects to Overseas Talents," 2004.

50. Gilley 1999.

51. Simons and Zielenziger 1994, p. 13A.

52. U.S. Embassy, Beijing, 1997.

53. Interview conducted by Guoqiang Yu with Andy Lee, August 14, 1999.

54. Lee now works in the software business that he envisioned, but it remains under the leadership of top officials of the Academy of Science.

55. Wu 2005b.

56. Wu 2005a.

57. Vendors justify expansion of the PAS system by describing it as a fixed-line, limited wireless solution (rather than a wireless network, for which they lack a license). The MII has so far turned a blind eye to the practice.

58. "China Sees Growing Tide," 2002.

59. A 2005 survey of the membership of the Chinese American Semiconductor Professionals Association (CASPA) conducted by Hubert Chen and Jian-Hung Chen found that Shanghai ranked first in perceived career potential, followed by the United States, Taiwan, and Beijing, with the Pearl River Delta/Hong Kong ranked lowest. They also ranked growth potential as the most significant determinant of decisions to relocate, ahead of other factors such as family or quality of life.

60. Zweig 2004.

61. Kirby 2002, p. B4.

62. Kirby 2002.

63. Wang and Wang 2002.

64. "Economic Analysis," 2004.

65. Dedrick 2003.

66. Xia 2004; "Chinese Students Return," 2003.

67. "China's Economy: Who Owns It?" 2004.

68. Studwell 2002.

69. A 1999 survey by the World Bank's International Finance Corporation reports that 80 percent of Chinese private sector companies believe the lack of access to financing has constrained their growth. Cited in Studwell (2002).

70. White et al. 2002.

71. Interview with Julie Yu Li, January 11, 2001.

72. Interview with entrepreneur who requested anonymity; April 16, 2001.

73. Xiao 2002.

74. CIEC 2001.

75. Xiao 2002.

76. Interview with Tang Hui-hao, September 10, 2004.

77. Xiao 2002.

78. "China's Champions," 2005; Wank 1999.

79. Saxenian and Quan 2005.

80. Interview with Haigeng Wang, January 7, 2001.

81. Saxenian and Quan 2005.

82. Ibid.

83. Kennedy 2005.

84. Wang and Wang 1998.

85. Howell et al. 2003, p. 109.

86. Segal 2003.

87. Ibid., p. 97. For example, the Shanghai government is now building a $12 billion deep water port with extra tunnels and a new bridge. It has also announced a $2.5 billion 36-kilometer bridge linking Shanghai and Ningbo in Zhejiang province (across the Hanzhou Bay), which will be the longest bridge spanning water in the world.

88. Segal 2003, p. 113. Howell and colleagues (2003, p. 50) summarize Circular 54 as follows: "For example, the construction of IC manufacturing facilities by even nominally private companies is granted the status of a 'government project' and is to be provided explicit loan subsidies, and the IC industry is given its own 24-hour 'window' at customs for '10-hour-or-less' service on all imports and exports; all IC chips manufacturing projects will, within three years after the date of accreditation, be exempted from the pay-

ment of handling charges and property right registration fees in purchasing housing for production and management use, and will be exempted from the payment of project-related water supply capacity increase charges, gas supply capacity increase charges, and power supply and distribution service charges."

89. Interview with Hua Ming, June 2, 2001.

90. Chwang 2003.

91. Gary Rivlin, "Talk of a Bubble as Venture Capitalists Flock to China," *New York Times,* December 6, 2004, p. C10.

92. The *xiaolingtong* also appeals to local government officials who have very modest salaries, and to local telecommunications carriers because the PAS system connects easily to legacy networks, and installation is very inexpensive. These constituencies likely account for the MII's unwillingness to clamp down on UTStarcom when it initially deployed the PAS system in Beijing and Shanghai.

93. Linn 1998.

94. Valigra 1998.

95. UTStarcom's revenues reflected the growth of unit sales, from $369 million in 2000 to $982 million in 2002 and $2.7 billion in 2004. Gross margins remained over 30 percent through 2003, but they fell in 2004, however, to 22 percent, reflecting intensifying competition as well as the saturation of their core market. UTStarcom Investor Presentation by Mike Sophie, Chief Operating Officer, 2004. Available online: *http://investorrelations.utstar.com/downloads/presentation.pdf.*

96. Clendenin 2003b.

97. Logan 2004.

98. Ellen Lee, "UTStarcom Eyes Other Markets for Expansion," *Knight-Ridder Tribune Business News,* September 18, 2005.

99. In 2003, 98 percent of IC designs in China had transistor counts of 5 million or fewer, and 72 percent incorporated 20,000 or fewer transistors. Only 6 percent of IC designs had clock frequency of 500 MHz or more. Data in this paragraph are taken from PricewaterhouseCoopers (2004).

100. De Filippo et al. 2005. They also differ substantially in market orientation: while outsourcing to foreign customers accounts for over 70 per-

cent of India's software, it accounts for only 10 percent of China's (mainly to Japanese customers).

101. De Filippo et al. 2005; "Outsourcing: Make Way for China," 2003.

102. "CEO Roundtable: Electronics Industry's Movers and Shakers of 2003," *Electronics Business,* August 15, 2003, p. 55.

103. Hua 2004.

104. Interview with Min Zhu, April 16, 2001.

105. WebEx Communications 10K Report to the U.S. Securities and Exchange Commission, 2004. In 2004 WebEx acquired an Indian software engineering firm, Cyberbazaar, which was located in Bangalore. This likely reflected the connections provided by the firm's co-founder, Subrah Iyar, who holds a B.S. from the Indian Institute of Technology, Mumbai, and an M.S. in computer engineering from the University of Southwestern Indiana, and has worked in Silicon Valley for over twenty years.

When WebEx co-founder Min Zhu resigned from the company in 2005, he announced his intention to promote the development of cross-border business between Silicon Valley and China. One aspect of this vision was developing a technology park in Shanghai that would provide even small U.S. companies with western-style business infrastructure that would allow them to succeed in China.

106. Interview with Larry Chen, December 3, 2003.

107. Lemon 2002.

108. Hua 2004.

109. Unicode provides an international standard for encoding the text of any document to be stored by a computer (this includes scripts, symbols, and character sets that are especially problematic in multilingual environments). The Unicode scheme assigns a unique number for every character, no matter what the platform, the program, or the language. It significantly reduces the challenges associated with international software development projects and has been widely adopted. See *www.unicode.org* for more information.

110. Saxenian and Quan 2005.

111. Ibid.

112. Naughton and Segal 2001, p. 38.

113. In December 2002, Neusoft became the first Chinese enterprise certified at CMM Level 5; SRS2 is certified at Level 4, and Achievo at Level 3.

114. Interview with Victor Wang, July 7, 1997.

115. Woetzel 2004.

116. For example, SMIC's early customers were overwhelmingly outside of China: according to the company's 2004 10-K filing, in 2003 36.7 percent were in North America, 26.7 percent in Taiwan, 12.5 percent in South Korea, 11.2 percent in Japan, 11 percent in Europe, and 1.9 percent in the rest of Asia-Pacific, including China.

117. Quan 2005.

118. Saxenian and Quan 2005.

119. Wu 2004.

120. Interview with Min Zhu, April 16, 2001.

121. *China Economics Quarterly* 2004; "Help Wanted," 2004.

7. IT ENCLAVES IN INDIA

1. Interview with Sanjay Anandaram, July 3, 1997.

2. NASSCOM 2000.

3. United Nations Development Programme 2003.

4. "The Kohli Touch," 2000.

5. Desai 2003.

6. Solomon and Kranhold 2005.

7. Madhavan 2005.

8. For a full account of Radha Basu's work on behalf of HP, see Goswamy (2003).

9. Goswamy 2003.

10. Parthasarathy 2000. Parthasarathy cites an interview with N. Seshagiri, former Additional Secretary, Department of Electronics, New Delhi, June 24, 1996.

11. Goswamy 2003.

12. Ibid.

13. Interview with Ashok Khosla, May 25, 1997.

14. Interview with Radha Basu, October 1, 1997.

15. Interview with Nimish Mehta, July 16, 1997.

16. Goswamy 2003, p. 84.

17. Ibid.

18. Ibid., p. 101.

19. Krishnadas 2004.

20. Interview with N. R. Narayana Murthy, December 11, 1997. Arthreye (2005) similarly concludes that tight labor market conditions were an important stimulus for investment in process capabilities by Indian software firms.

21. "A World of Work," 2004, p. 10.

22. Interview with Narayana Murthy, December 11, 1997.

23. Cited in Kumra and Sinha (2003).

24. *Dataquest* 2002.

25. Mohnot 2003.

26. De Filippo et al. 2005.

27. Sen and Frankel 2005.

28. Bahl 2005.

29. Stremlau 1996, p. 157.

30. Parthasarathy 2004.

31. Desai 2003.

32. Ibid.

33. Parthasarathy 2000.

34. Agrawal et al. 2003.

35. These and the following quotations from Raman Roy are taken from "The Little Start-Up That Could," 2004.

36. Farrell and Zainulbhai 2004, pp. 50–59.

37. All the data in this paragraph are drawn from Sen and Frankel (2005).

38. Statistics in this and the previous paragraph taken from Varma 2004 and Arora 2005.

39. Kumra and Sinha 2003.

40. Bahl 2005.

41. Hamm 2005.

42. Kurian 1997.

43. Schram 2001, p. 66.

44. Roy 2004.

45. Biswas 1998.

46. Desai 2003.

47. Krishnadas 1997.

48. Ibid.

49. Bruce Sterling, "Simputer," *New York Times,* December 9, 2001; Mike Langberg, "Simputer Aimed at Easing Life in Indian Villages," *The Mercury News,* August 14, 2002. For more coverage, see *www.ncoretech.com/simputer/newsroom/index.html.*

50. Desai 2003.

51. Mathur 2002.

52. Natarjun 2005.

53. Santa Clara County Office of Human Relations, 2000.

54. Wilson 2004.

55. Srinivasan 2005.

56. See the *Report of K. B. Chandrasekhar Committee on Venture Capital* to the Securities and Exchange Board of India at *www.sebi.gov.in.* The SEBI board adopted the report in January 2000, signaling the seriousness of the government's intention to pursue its recommendations.

57. Biswas 1998.

58. Anandaram 2003.

59. Ashok 2004.

60. Shankar and Sundaram 2003.

61. Roberts 2004.

62. Sundaram 2003, p. 23.

63. Krishnadas 2005.

64. Jayashankar 2004.

65. Pete Engardio, "Scouring the Planet for Braniacs," *Business Week,* October 11, 2004.

66. Mitta 2004.

67. Roberts 2004.

68. Chiruvolu 2003.

69. Roberts 2004.

70. Roy 2004.

71. "It was the right time to go," interview with Vivek Paul, *rediff.com,*

July 8, 2005. Available online: *www.rediff.com/money/2005/jul/08inter.htm.* Access date: September 12, 2005.

72. For more details see "Now, a mobile desktop at Rs 10,000," *The Economic Times,* May 10, 2005; "Now, mobile desktop PCs!" *rediff.com,* May 10, 2005; and G. Ananthahkrishnan, "Low cost computer maker to work with software community," *The Hindu*, May 19, 2005.

73. Farrell and Lund 2005.

74. Srinivasan 2003.

75. *www.internetworldstats.com*

76. Neelakantan 2005.

77. Vijaybaskar et al. 2000.

78. *China Economics Quarterly* 2004.

79. Paul 2005.

8. THE ARGONAUT ADVANTAGE

1. Helper, MacDuffie, and Sabel (2000) refer to these processes as "pragmatist disciplines" because they "oblige firms routinely to question the suitability of their current routines and continuously to readjust their ends and means to one another in light of the results of such questioning."

2. Two experienced engineers summarize recent research on global software development: "As we increasingly work in virtual, distributed team environments, we will more and more face formidable problems of miscommunication, lack of coordination, infrastructure compatibility, cultural misunderstanding, and conflicting expectations—not to mention architecting products for distributed development" (Herbsleb and Moitra 2001).

3. Olson and Olson (2000) conclude that effective distance work depends on high common ground, loose coupling of work, and readiness for collaboration and collaboration technologies.

4. Even application software must be developed in conjunction with the business process it automates.

5. Trajtenberg 2000.

6. Mahmood and Singh 2003.

7. Rasiah 2001; McKendrick et al.2000.

8. Population: Trajtenberg 2000. Exports: Mahmood and Singh 2003.

9. Mahmood and Singh 2003.

10. Two recent studies conclude that job losses due to offshoring have been small relative to the general job turnover in industrialized economies. See Bednarzik 2005; OECD 2004.

11. Research and development (R&D) in both China and India remains more "D" than "R." See Walsh 2003 and Desai 2003.

12. This and the following two quotations from Eric Benhamou are taken from "Electronics Industry's Movers & Shakers," 2003.

REFERENCES

Addison, Craig. 2001. *Silicon Shield: Taiwan's Protection Against Chinese Attack.* Irving, Tex.: Fusion Press.

Agrawal, Vivek, Diana Farrell, and Jaana K. Remes. 2003. "Offshoring and Beyond." *McKinsey Quarterly* 4: 24–35.

Amsden, Alice H. 1989. *Asia's Next Giant: South Korea and Late Industrialization.* New York: Oxford University Press.

———. 2001. *The Rise of "the Rest": Challenges to the West from Late-Industrializing Economies.* New York: Oxford University Press.

Amsden, Alice H., and Wan-wen Chu. 2003. *Beyond Late Development: Taiwan's Upgrading Policies.* Cambridge, Mass.: MIT Press.

Anandaram, Sanjay. 2003. "The Early Stage U.S.-India Cross-Border Technology Start-up." *Asian Venture Capital Journal,* September 29. Available online: *http://www.asianfn.com/* (access by password only).

Angwin, Julia. 1997. "The Israeli Connection: Silicon Valley Promised Land for Startups." *San Francisco Chronicle,* September 10. Available online: *http://sfgate.com/.* Access date: September 2, 2004.

Arora, Ashish, and Alfonso Gambardella, eds. 2005. *From Underdogs to Tigers.* New York: Oxford University Press.

Arora, Ashish, et al. 2001. "The Indian Software Services Industry." *Research Policy* 30 (8, October): 1267–1287.

Arora, Shipra. 2005. "The Indian Multinational." *Dataquest India,* May

9, 2005. Available online: *http://dqindia.com*. Access date: May 5, 2005.

Arthreye, Suma S. 2005. "The Indian Software Industry and Its Evolving Service Capability." *Industrial and Corporate Change* 14 (3): 365–391.

Arthur, W. Brian. 1994. *Increasing Returns and Path Dependence in the Economy.* Ann Arbor: University of Michigan Press.

ASE Inc., Kaohsiung. 20-F filing, 2002. Available online: *http://www.asetwn.com.tw/*.

Ashkenas, Ron, Dave Ulrich, Todd Jick, and Steve Kerr. 2003. *The Boundaryless Organization: Breaking the Chains of Organizational Structure.* San Francisco: Jossey-Bass.

Ashok, Sathya Mithra. 2004. "Seeking Silicon Valley," *CyberIndia Online,* March 5. Available online: *http://www.ciol.com/*. Access date: September 2, 2004.

Asian Americans for Community Involvement. 1993. Available online: *http://www.aaci.org/index.htm*.

Autler, Gerald. 2000. "Territorially-Based Learning in a Global Economy: The Semiconductor Industry in Israel." Master's Thesis, Department of City and Regional Planning, University of California, Berkeley.

———. 2002. "Territorially-Based Learning in a Global Economy: The Semiconductor Industry in Israel." Unpublished paper, Department of City and Regional Planning, University of California, Berkeley.

"A World of Work: A Survey of Outsourcing." 2004. *The Economist,* November 13, 2004.

Bahl, Sheetal. 2005. "Is Offshoring Demand Sustainable?" Everest Research Institute, July. Available online: *www.outsourcing–journal.com/*. Access date: July 8, 2005.

Baijia, Lui. 2003. "IC Professionals Badly Needed." *China Daily,* February 10. Available online: *http://www.chinadaily.com*. Access date: September 2, 2004.

Baldwin, Carliss Y., and Kim B. Clark. 2000. *Design Rules: The Power of Modularity.* Cambridge, Mass.: MIT Press.

Bardhan, Ashok Deo, and David K. Howe. 1998. "Transnational Social

Networks and Globalization: The Geography of California's Exports." Working Paper No. 98-262, February 1. Berkeley: Fisher Center for Real Estate and Urban Economics, University of California, Berkeley.

Barrett, Richard, and Martin K. Whyte. 1982. "Dependency Theory and Taiwan: Analysis of a Deviant Case." *American Journal of Sociology* 87 (5, March): 1064–1089.

Bartlett, Christopher A., and Sumantra Ghoshal. 1998. *Managing Across Borders: The Transnational Solution.* Boston: Harvard Business School Press.

Bednarzik, Robert. 2005. "Restructuring Information Technology: Is Offshoring a Concern?" *Monthly Labor Review.* Washington, D.C.: Bureau of Labor Statistics.

Best, Michael. 1990. *The New Competition: Institutions of Industrial Restructuring.* Cambridge, Mass.: Harvard University Press.

———. 2001. *The New Competitive Advantage: The Renewal of American Industry.* New York: Oxford University Press.

Bhagwati, Jagdish N. 1985. *Essays in Development Economics,* ed. Gene Grossman. Cambridge, Mass.: MIT Press.

Biswas, Smita. 1998. "Leading the VC Wave in India." *siliconindia,* May/June. *http://web.archive.org/.* Access date: September 19, 2004.

Blackburn, Nicky. 2003. "ICQ Founder Wasn't Seeking Fortune." *Jerusalem Post,* July 10, p. 12.

Blaug, Mark. 1986. *Economic History and the History of Economics.* New York: New York University Press.

Bordo, Michael, Alan Taylor, and Jeffrey Williamson, eds. 2003. *Globalization in Historical Perspective.* Chicago: University of Chicago Press.

Borrus, Michael G. 1988. *Competing for Control: America's Stake in Microelectronics.* Cambridge, Mass.: Ballinger.

Borrus, Michael G., D. Ernst, and S. Haggard, eds. 2000. *International Production Networks in Asia: Rivalry or Riches.* London: Routledge.

Brenner, Reuven. 1994. *Labyrinths of Prosperity: Economic Follies, Democratic Remedies.* Ann Arbor: University of Michigan Press.

Breznitz, Danny. 2000. "High Tech, Innovation, and the Periphery: The Role of the Military in the Israeli Success." *PreCIS* 10(4): 8–10.

Brown, John Seeley, and Paul Duguid. 2000. *The Social Life of Information.* Boston: Harvard Business School Press.

Brusco, Sebastian. 1982. "The Emilian Model: Productive Decentralization and Social Integration." *Cambridge Journal of Economics* 6 (2, June): 167–184.

Bureau of Standards, Metrology & Inspection (BSMI). 2000. "The Accumulated Number of IS0–9000 Certified Companies." Ministry of Economic Affairs, ROC.

Burton, Lawrence, and Jack Wang. 1999. "Issue Brief: How Much Does the U.S. Rely on Immigrant Engineers?" Division of Science Resource Studies, National Science Foundation, Arlington, Va. February 11.

Callon, Scott. 1995. "Different Paths: The Rise of Taiwan and Singapore in the Global Personal Computer Industry." Asia/Pacific Research Center, Stanford University, January.

Chan, Wai-Chan, Martin Hirt, and Stephen Shaw. 2002. "The Greater China High-Tech Highway." *McKinsey Quarterly* 4. Available online: *http://www.mckinseyquarterly.com/.* Access date: September 18, 2004.

Chandler, Alfred D., Jr. 1977. *The Visible Hand: The Managerial Revolution in American Business.* Cambridge, Mass.: Harvard University Press.

Chandler, Alfred D., Jr., and Herman Daems, eds. 1980. *Managerial Hierarchies: Comparative Perspectives on the Rise of the Modern Industrial Enterprise.* Cambridge, Mass.: Harvard University Press.

Chang, Pao-Long, Chintay Shih, and Chiung-Wen Hsu. 1994. "The Formation Process of Taiwan's IC Industry—Method of Technology Transfer." *Technovation* 14(3): 161–171.

Chang, Pao-Long, Chiung-Wen Hsu, and Chien-Tzu Tsai. 1999. "A Stage Approach for Industrial Technology Development and Implementation—The Case of Taiwan's Computer Industry." *Technovation* 19: 233–241.

Chang, Pao-Long, and Chiung-Wen Hsu. 2001. "The Industrial Park: Government's Gift to Industrial Development." In Chun-Yen Chang and Po-Lung Yu, eds., *Made in Taiwan: Booming in the Information Era.* Hackensack, N.J.: World Scientific.

Chang, Shirley L. 1992. "Causes of Brain Drain and Solutions: The Taiwan Experience." *Studies in Comparative International Development* 27(1): 27–43.

Chen, An-Pin, and Shinn-Wen Wang. 2001. "Employee Profit Sharing and Stock Ownership Attracts World-Class Employees." In Chun-Yen Chang and Po-Lung Yu, eds., *Made in Taiwan: Booming in the Information Era.* Hackensack, N.J.: World Scientific.

Chen, Kathy. 2004. "China Will Keep Pursuing Digital Standards." *Wall Street Journal,* April 23, 2004, pp. B1–B2(4).

Chen, Ming-chi. 2002. "Industrial District and Social Capital in Taiwan's Economic Development: An Economic Sociological Study on Taiwan's Bicycle Industry." Ph.D. dissertation, Department of Sociology, Yale University.

Chen, Shin-Horng. 2002. "Global Production Networks and Information Technology: The Case of Taiwan." *Industry and Innovation* 9 (3): 249–265.

Chen, Tain-Jy. 2003. "Network Resources for Internationalization: The Case of Taiwan's Electronics Firms." *Journal of Management Studies* 40 (5): 1107–1125.

Cheng, Lu-lin. 1996. "Embedded Competitiveness: Taiwan's Shifting Role in the International Footwear Sourcing Networks." Ph.D. dissertation, Department of Sociology, Duke University.

Cheng, Lu-lin, and Gary Gereffi. 1994. "The Informal Economy in East Asian Development." *International Journal of Urban and Regional Research* 18(2): 194–219.

Chesbrough, Henry W. 2003. *Open Innovation: The New Imperative for Creating and Profiting from Technology.* Boston: Harvard Business School Press.

China Economics Quarterly. 2004. "The Economics of Education in China." *FT.com,* October 11, 2004. Available online: *http//news.ft.com/cms.* Access date: December 6, 2004.

"China Sees Growing Tide of Returning Talent." 2002. *Xinhua News Agency,* June 28.

China Statistics Press. 2003. *China Statistics Yearbook on High Technology Industry.* Beijing: China Statistics Press.

"China's Champions: Special Report." *The Economist,* January 8, 2005, pp. 59–61.

"China's Economy: Who Owns It? Changes in Ownership Patterns in China's Economy, 1992–2002." 2004. A report by the *China Economics Quarterly,* Edition 1. May. Hong King: Dragonomics Ltd.

Chinese Institute of Engineers–USA. 1999. "CIE-USA Annual Awards, 1950s–1990s." Available online: *http://www.cie-gync.org/.* Access date: July 27, 1999.

"Chinese Students Return from Abroad to Success and Failure." 2003. *Xinhua News Agency,* December 17.

Chinitz, Benjamin. 1961. "Contrasts in Agglomeration: New York and Pittsburgh." *Papers and Proceedings of the American Economic Association,* May.

Chiruvolu, Ravi. 2003. "TechTalk: About That India Recommendation . . ." *Venture Capital Journal,* November 1. Available online: *http://www.venturecapitaljournal.net/.* Access date: September 19, 2004.

Chugani, Michale. 2002. "Chinese IT Professionals Flock Back Home." *South China Morning Post, January 12.* Available online: *http://www.vuw.ac.nz/.* Access date: September 22, 2004.

Chwang, Ronald. 1997. Welcome and Opening Remarks, Monte Jade Science and Technology Association (West). Seventh Annual Conference, Santa Clara, Calif., March 15.

———. 2003. "Venture Investment in Greater China—Opportunities and Challenges." Presentation at Stanford University, May 29.

Clendenin, Mike. 2003a. "Befriending the Dragon." *Electronic Engineering Times,* October 31. Available online: *http://www.eetimes.com/.* Access date: September 2, 2004.

———. 2003b. "Comlent Brings High-End RF to China." *Electronic Business News,* November 17. Available online: *http://www.ebnonline.com/.* Access date: September 6, 2004.

———. 2003c. "Comment: For SMIC, Rumblings of an IPO." *EETimes,* August 1. Available online: *http://www.siliconstrategies.com/.* Access date: August 21, 2004.

———. 2004. "Creative Play All in a Day's Work at ITRI." *Electronic Engineering Times,* August 16. Available online: *www.eetimes.com/print/archive.* Access date: August 28, 2004.

Cohen, Y., and Y. Haberfeld. 2001. "Self Selection and Return Migration: Israeli-Born Jews Returning Home from the United States During the 1980s." *Population Studies* 55: 79–91.

Cole, Robert. 2003. "Japanese Manufacturing Dilemmas: The ICT Industries." Paper presented at Conference on Institutional Change in East Asia, Cornell University, April 3–5.

Committee of 100. 2001. "American Attitudes Toward Chinese Americans and Asian Americans." A White Paper prepared for the Committee of 100 by Yankelovich Partners. New York, April 2001.

Common Wealth. 1996. July 1, 1996.

Cooke, Merritt T. 2004. "The Politics of Greater China's Integration into the Global Info Tech (IT) Supply Chain." *Journal of Contemporary China* 13(40): 491–506.

Cortright, Joseph. 2005. "Making Sense of Clusters." Draft prepared for The Brookings Institution Metropolitan Policy Program.

Dataquest. 2002. "Dataquest 20 Years: The Great Indian Software Revolution." Available online: *www.dqindia.com//content/.* Access date: December 23, 2002.

Dedrick, Jason. 2003. Paper presented at Conference on the Global IT Industry: The Future of China and India. University of California, Santa Cruz, May 30, 2003.

Dedrick, Jason, and Kenneth Kraemer. 1998. *Asia's Computer Challenge: Threat or Opportunity for the United States and the World?* New York: Oxford University Press.

De Filippo, Giuseppe, Jun Hou, and Christopher Ip. 2005. "Can China Compete in IT Services?" *McKinsey Quarterly,* No. 1.

deFontenay, Catherine, and Erran Carmel. 2004. "Israel's Silicon Wadi: The Forces behind Cluster Formation." In T. Bresnahan and A. Gambardella, eds., *Building High-Tech Clusters: Silicon Valley and Beyond.* Cambridge: Cambridge University Press.

Desai, Ashok V. 2003. "The Dynamics of the Indian Information Technology Industry." Center for New and Emerging Markets, London Business School, April. Available online: *http://www.london.edu/.* Access date: September 6, 2004.

Downes, Larry. 2002. *The Strategy Machine: Building Your Business One Idea At a Time.* New York: HarperBusiness.

"Economic Analysis of Zhangjiang Hi-Tech Park 2003." 2004. *Park News,* February 18. Available online: *www.zjpark.com/.* Access date: August 21, 2004.

Ehrgood, Thomas. 2000. Presentation by International Trade Counsel, Compaq Computer Corporation, at Chairman's Circle Luncheon, 2000 Annual Joint Business Conference, The Grand Hotel, Taipei, Taiwan, June 14. Available online: *http://www.usa-roc.org/reports/.* Access date: March 14, 2001.

Electronic Business CEO Roundtable. 2003. "Electronics Industry's Movers & Shakers." Available online: *http://www.reed-electronics.com/.* Access date: September 6, 2004.

Engardio, Pete. 1993. "Taiwan: 'The Arms Dealer of the Computer Wars.'" *Business Week,* June 28, pp. 51–53.

Ernst, Dieter. 1994. "What Are the Limits to the Korean Model? The Korean Electronics Industry under Pressure." Research paper, Berkeley Roundtable on the International Economy. University of California, Berkeley.

———. 2000. "Carriers of Cross Border Knowledge Diffusion: Information Technology and Global Production Networks." East-West Center Working Papers, Economics Series, No. 3. Honolulu: East-West Center.

———. 2003. "Internationalisation of Innovation: Why Is Chip Design Moving to Asia?" East-West Center Working Papers, Economics Series, No. 64. Honolulu: East-West Center.

Ernst, Dieter, and Linsu Kim. 2002. "Global Production Networks, Knowledge Diffusion, and Local Capability Formation." *Research Policy* 31 (8–9, December): 1417–1429.

Espenshade, Thomas J., Margaret L. Usdansky, and Chang Y. Chung. 2001. "Employment and Earnings of Foreign-Born Scientists and Engineers." *Population Research and Policy Review* 20: 81–105.

Fallick, Bruce, Charles A. Fleischman, and James B. Rebitzer. 2003. "Job Hopping in Silicon Valley: The Micro-Foundations of a High Technology Cluster." Working Paper, Weatherhead School of Management, Case Western Reserve University. Available online: *http://wsomfaculty.cwru.edu/rebitzer/.* Access date: September 16, 2005.

Farrell, Diana, and Adil S. Zainulbhai. 2004. "A Richer Future for India." *McKinsey Quarterly,* special edition: 50–59.

Farrell, Diana, and Susan Lund. 2005. "Reforming India's Financial System." *The McKinsey Quarterly,* 2005 Special Edition. Available online: *www.mckinseyquarterly.com.* Access date: September 26, 2005.

Ferguson, Charles H. 1990. "Computers and the Coming of the U.S. Keiretsu." *Harvard Business Review* 4: 55–70.

Flannery, Russell. 1999. "Taiwan's PC Production Spills into China." *The Wall Street Journal,* October 21, p. A16.

"Former CASPA Chairmen Took New Career Journeys in Shanghai." 2001. *Sing Tao Daily,* October 8. Available online: *http://www.caspa.com/newsletter/.* Access date: September 23, 2004.

Galbraith, John K. 1967. *The New Industrial State.* Boston: Houghton Mifflin.

Gereffi, Gary, and Miguel Korzeniewicz, eds. 1994. *Commodity Chains and Global Capitalism.* Westport, Conn.: Praeger.

Gerschenkron, Alexander. 1962. *Economic Backwardness in Historical Perspective: A Book of Essays.* Cambridge, Mass.: Harvard University Press.

Gilley, Bruce. 1999. "Looking Homeward." *Far Eastern Economic Review,* March 11. Available online: *http://www.feer.com/articles/1999/.* Access date: September 6, 2004.

Gillmor, Dan. 1998. "High-Tech Blooming in Israel." *San Francisco Chronicle,* April 27.

Gilson, Ron J. 1998. "The Legal Infrastructure of High Technology Industrial Districts: Silicon Valley, Route 128, and Covenants Not to Compete." Stanford Law School, John M. Olin Program in Law and Economics, Working Paper No. 163, August.

Global Insight, Inc. 2004. *Venture Impact 2004.* Available online: *http://www.nvca.org/.* Access date: September 22, 2004.

Goering, Richard. 2004. "Design Tools 'Made Not in USA.'" *Electronic*

Engineering Times, March 29. Available online: *http://www.eet.com/.* Access date: September 6, 2004.

Gold, Thomas B. 1986. *State and Society in the Taiwan Miracle.* Armonk, N.Y.: Sharpe.

Gompers, Paul, Josh Lerner, and David Scharfstein. 2003. "Entrepreneurial Spawning: Public Corporations and the Genesis of New Ventures, 1986–1999." Working Paper 9816, National Bureau of Economic Research. Available online: *http://www.nber.org/papers/.* July. Access date: September 6, 2004.

Goswamy, Kavita. 2003. "The Seventh Avenue of Return Contribution: Expatriate Indian Executives Championing India's Information Technology." Undergraduate honors thesis, University of California, Berkeley.

Gramsci, Antonio. 1992. *The Prison Notebooks.* New York: Columbia University Press.

Granovetter, Mark. 1995. "The Economic Sociology of Firms and Entrepreneurs." In A. Portes, ed., *The Economic Sociology of Immigration.* New York: Russell Sage.

Guan, Zhixiong. 2002. "Overcoming the China Syndrome in Japan." Forum on Integration of the Asian Economies, Present and Future, Tokyo, Japan, April 22–23.

Guochu, A., and L. Wenjun. 2001. "International Mobility of China's Resources in Science and Technology and Its Impact." Report number COM/DTSI/DEELSA/RD. Paris: OECD.

Hagel, John, III. 2004. "Offshoring Goes on the Offensive." *McKinsey Quarterly* 2. Available online: *http://www.mckinseyquarterly.com/.* Access date: September 19, 2004.

Hall, Chris. 1999. "Leader and Dealer: AsusTek Computer." *Electronic Business Asia,* October. Available online: *www.eb-asia.com/.* Access date: August 21, 2004.

———. 2000. "Living in a Sub-$1000 World." *Electronic Business Asia,* February. Available online: *www.eb-asia.com/.* Access date: August 21, 2004.

Hamilton, Gary. 1997. "Organization and Market Processes in Taiwan's

Capitalist Economy." In Marco Orru et al., eds., *The Economic Organization of East Asian Capitalism.* Thousand Oaks, Calif.: Sage.

Hamilton, Ian. 2005. "Parsing the 'China Price.'" *Electronic Engineering Times,* January 25, 2005. Available online: *www.eetimes.com/article.* Access date: January 29, 2005.

Hamm, Steve. 2005. "Taking a Page from Toyota's Playbook." *Business Week,* August 22, 2005.

Hellman, Thomas. 1998. "WI Harper International: Bridge Between Silicon Valley and Asia." Graduate School of Business, Stanford University, SM-39.

Helper, Susan, John Paul MacDuffie, and Charles Sabel. 2000. "Pragmatic Collaborations: Advancing Knowledge While Controlling Opportunism." *Industrial and Corporate Change,* vol. 10, no. 3: 443–483.

"Help Wanted." 2004. *The Economist,* October 7. Available online: *www.economist.com/world/asia.* Access date: October 8, 2004.

Herbsleb, James D., and Deependra Moitra. 2001. "Global Software Development." *IEEE Software* (March/April): 16–20.

Herrigel, Gary. 1996. *Industrial Constructions: The Sources of German Industrial Power.* Cambridge: Cambridge University Press.

Hill, Susan T. 1996. "Data Brief: Non U.S. Citizens are 40 Percent of S&E Doctorate Recipients from U.S. Universities in 1995." Vol. 1996, No. 9. Washington, D.C.: Division of Science Resources Studies, National Science Foundation, Arlington, Va. August 19.

Hobday, Michael. 1995. "East Asian Latecomer Firms: Learning the Technology of Electronics." *World Development* 23 (7): 1171–1193.

———. 2001. "The Electronics Industries of the Asia-Pacific: Exploiting International Production Networks for Economic Development." *Asian-Pacific Economic Literature* 15 (1, May): 13–29.

Howell, Thomas R., Brent L. Bartlett, William A. Noellert, and Rachel Howe. 2003. "China's Emerging Semiconductor Industry: The Impact of China's Preferential Value-Added Tax on Current Investment Trends." Paper prepared by Dewey Ballantine LLP for the Semiconductor Industry Association, October.

Hsu, Jinn-Yuh. 1997. "A Late Industrial District? Learning Networks in the Hsinchu Science-Based Industrial Park." Ph.D. dissertation, Department of Geography, University of California, Berkeley.

Hua, Vanessa. 2004. "Looking Offshore: China. Giving India Competition, Chinese-Born Entrepreneurs Help Homeland." *SFGate.com,* March 7. Available online: *http://sfgate.com/*. Access date: September 6, 2004.

IMD International. 2004. *IMD World Competitiveness Yearbook 2004.* Lausanne, Switzerland: IMD International.

Israel Venture Capital (IVC) Research Center. 2004. "Most Active Foreign Investors in Israel, Survey." Available online: *www.ivc-online.com.* Access date: July 13, 2004.

Jayashankar, Mitu. 2004. "Ittiam Systems: The 3rd Option." *Businessworld,* July 12. Available online: *www.businessworldindia.com/*. Access date: May 15, 2005.

Jhunjhunwala, Ashok, et al. 1998. "The Role of Technology in Telecom Expansion in India." *IEEE Communications Magazine,* December. Available online: *http://www.tenet.res.in/jhun/papers/Telxpan/telxpan.html.* Access date: September 13, 2004.

Jiang, Jim. 2001. "We Stay in Touch and They Often Ask About China and the Market and Admire Our Achievements and Want to Return." Netbig.com interview, January 10. Available online: *http://Netbig.com.*

Jiang, Xiaodong, Lynn Lee, Qiaowei Shen, and Sarah Wang. 2004. "Development of Industrial Clusters in China's Electronics Industry." UCB-UNIDO Research Project.

Joint Venture: Silicon Valley Network. 2004. *The 2004 Index of Silicon Valley.* San Jose, Calif. Available online: *http://www.jointventure.org/*. Access date: September 19, 2004.

Kahn, Alan. 2004. "Profiting from Benchmarks: Reasons Behind China's Rush to Develop Standards." *www.chinatechnews.com,* April 7. Access date: April 20, 2004.

Kennedy, Scott. *The Business of Lobbying in China.* Cambridge, Mass.: Harvard University Press, 2005.

Khavul, Susanna. 2003. "The Emergence and Evolution of Israel's Software

Industry." Center for New and Emerging Markets, London Business School, June.

Kirby, Carrie. 2002. "Tech Professionals Returning to China." *San Francisco Chronicle,* January 2, pp. B1–B4.

———. 2003. "Opportunities in China Entice Overseas Chinese." *The San Francisco Chronicle,* January 12, p. B1.

Kishimoto, Chikashi. 2003. "Upgrading in the Taiwanese Computer Cluster: Transformation of Its Production and Knowledge Systems." Working Paper 186. Sussex, England: Institute of Development Studies.

Kogut, Bruce, and Udo Zander. 1993. "Knowledge of the Firm and the Evolutionary Theory of the Multinational Corporation." *Journal of International Business Studies* 24(4): 625–645.

Krishnadas, K. C. 1997. "Chops & Bytes: This Yantra Needs a Mantra." *The Economic Times (India),* March 25.

———. 2004. "Indian Engineers Repatriate." *Electronic Engineering Times,* March 26. Available online: *www.eet.com/.* Access date: September 6, 2004.

———. 2005. "Waiting for Ex-pats." January 10. Available online: *www.eetimes.com/news/.* Access date: May 15, 2005.

Krugman, Paul R. 1991. *Geography and Trade.* Cambridge, Mass.: MIT Press.

Kumra, Gautam, and Jayant Sinha. 2003. "The Next Hurdle for Indian IT." *The McKinsey Quarterly,* No. 4.

Kurian, Thomas. 1997. "In the Footsteps of Giants." *siliconindia,* November. Available online: *http://web.archive.org/web/.* Access date: September 19, 2004.

Lamoreaux, Naomi R., Daniel M. G. Raff, and Peter Temin. 2003. "Beyond Markets and Hierarchies: Toward a New Synthesis of American Business History." *The American Historical Review* 108, 2: 404–430.

Landler, Mark. 2001. "These Days 'Made in Taiwan' Often Means Made in China." *The New York Times,* May 29, pp. A1, C2.

Langois, Richard N. 2003. "The Vanishing Hand: The Changing Dynamics of Industrial Capitalism." *Industrial and Corporate Change* 12, 2: 351–385.

Lave, Jean. 1991. *Situated Learning: Legitimate Peripheral Participation.* Cambridge: Cambridge University Press.

Lee, Chuan-kai. 2005. "Cluster Adaptability Across Sector and Border." Ph.D. dissertation, Department of City and Regional Planning, University of California, Berkeley.

Lemon, Sumner. 2002. "Gartner: Look to China for Outsourcing Services." *CNN.com,* January 21. *http://www.cnn.com/.* Access date: September 6, 2004.

Lenoir, Timothy, N. Rosenberg, H. Rowen, C. Lecuyer, J. Colyvas, and B. Goldfarb. 2004. "Inventing the Entrepreneurial Region: Stanford and the Co-Evolution of Silicon Valley." Unpublished manuscript, Stanford University.

Leslie, Stuart, and Kargon. 1996. "Selling Silicon Valley: Frederick Terman's Model for Regional Advantage." *Business History Review* 70, 4: 435–472.

"Letter from Taiwan." 2002. *Electronic Business Asia,* January. Available online: *www.eb-asia.com/.* Access date: August 21, 2004.

Levy, Brian, and Wen-Jeng Kuo. 1991. "The Strategic Orientation of Firms and the Performance of Korea and Taiwan in Frontier Industries: Lessons from Comparative Case Studies of the Keyboard and Personal Computer Assembly." *World Development,* 19(4): 363–374.

Light, Ivan, and Edna Bonacich. 1988. *Immigrant Entrepreneurs: Koreans in Los Angeles, 1965–1982.* Berkeley: University of California Press.

Light, Ivan, and Elizabeth Roach. 1996. "Self-Employment in Southern California, 1970 to 1990." In Roger Waldinger and Mehdi Bozorgmehr, eds., *Ethnic Los Angeles.* New York: Russell Sage Foundation.

"Like Company Like Bank." 2000. *siliconindia,* July 24. Available online: *http://www.siliconindia.com/Magazine/.* Access date: September 19, 2004.

Lin, Tenglin. 2000. "The Social and Economic Origins of Technological Capacity: A Case Study of the Taiwanese Computer Industry." Ph.D. Dissertation, Department of Sociology, Temple University.

Lin, Xi-wei, Gang Bai, and Tim Wang. 1998. "China Semiconductor Industry Overview." Report on the first NACSA delegation to visit semiconductor R&D institutes. Available online: *http://www.nacsa.com/events/ChinaTrip.html.* Access date: August 21, 2004.

Linden, Greg. 1999. "Technology Policy and Growth: Case Studies from the Asian Electronics Industry." Ph.D. dissertation prospectus, University of California, Berkeley, February 7.

Linn, Gene. 1998. "Battling the Big Boys of Telecom." *Journal of Commerce,* May 27.

Logan, Michael. 2004. "Comlent Puts Its Chips in Niche Handsets." *South China Morning Post,* January 8. *http://www.comlent.com/.* Access date: September 23, 2004.

Luo, Yu-Ling, and Wang Wei-Jen. 2001. "High-Skill Migration and Chinese Taipei's Industrial Development." Seminar on International Mobility of Highly Skilled Workers: From Statistical Analysis to the Formulation of Policies, June 11–12. Working paper COM/DSTI/DEELSA/RD. Paris: OECD.

———. 2003. "Sunplus Partners with Silicon Image on Development of SoCs for Digital Media." DigiTimes.com, August 6. Available online: *http://www.digitimes.com/NewsShow/Article.asp.* Access date: September 27, 2004.

Ma, Debin. 2003. "Industrial Revolution in the Prewar Lower Yangzi Region of China: A Quantitative and Institutional Interpretation." Unpublished paper, Foundation for Advanced Studies in International Development, Tokyo, December 2003.

Madhavan, Narayanan. 2005. "Vivek Paul Put Wipro on World Map." Reuters. Available at: *www.ciol.com/content/news/.* Access date: July 1, 2005.

Maheshwari, Arun. 2005. "Indian Companies vs. Global MNCs." *Dataquest, www.dqindia.com.* Access date: May 9, 2005.

Mahmood, Ishtiaq P., and Jasjit Singh. 2003. "Technological Dynamism in Asia." *Research Policy* 32: 1031–1054.

Main Science & Technology Indicators. 1999. Organization for Economic Cooperation and Development. Available online: *http://puck.sourceoecd.org/.* Access date: September 23, 2004.

Mathews, John A. 1997. "A Silicon Valley of the East: Creating Taiwan's

Semiconductor Industry." *California Management Review* 39 (4, Summer): 26–54.

———. 2002. *Dragon Multinational: A New Model for Global Growth.* New York: Oxford University Press.

Mathur, Rakesh. 2002. Presentation at Forum on Local and Global Networks of Immigrant Professionals. San Francisco: Public Policy Institute of California.

McKendrick, David G., Richard F. Doner, and Stephan Haggard. 2000. *From Silicon Valley to Singapore: Location and Competitive Advantage in the Hard Disk Drive Industry.* Stanford: Stanford University Press.

Meany, Connie Squires. 1994. "State Policy and the Development of Taiwan's Semiconductor Industry." In J. Aberbach et al., *The Role of the State in Taiwan's Development.* London: Sharpe.

Minamikawa, A. "Investment in 300mm Plants Heating Up." 2000. *Nikkei Microdevices,* June (JPP20000706000024). Translated from *http://ne.nikkeibp.co.jp/NMD/.* Access date: September 15, 2004.

Mitta, Sridhar. 2004. "Intellectual Property Exchange Forum." Presentation at Electronics & Information Technology Exposition, New Delhi, April 26.

Mohnot, Navyug. 2003. "Why 'India Inside' Spells Quality." *Dataquest India,* October 27, 2003. Available online: *www.dqindia.com/.* Access date: November 1, 2004.

Moore, Jonathan, et al. 1998. "The Taiwan Touch." *Business Week,* May 25, p. 22.

Mufson, Steven. 1998. "The Information Highway—From Texas to Beijing." *Washington Post,* June 16, p. A1.

Murphy, David. 2001. "Holding the Purse Strings." *Far Eastern Economic Review,* July 5, 2001: 60–61.

NACSA 2002–2003 Annual Report. North America Chinese Semiconductor Association. Available online: *http://www.nacsa.com/.* Access date: September 22, 2004.

Naipaul, V. S. 1992. *India: A Million Mutinies Now.* New York: Penguin Books.

Nasa Tsai. 2002. "China: An Emerging Centre for Semiconductor Manufacturing." *Future Fab International* 13, July 8. Available online: *http://www.future-fab.com/*. Access date: September 6, 2004.

Natarjun, Arun. 2005. "Venture Intelligence India Roundup." TSJ Media, Arun Enterprises, January 10, 2005. Available online: *www.tsjmedia.com.*

National Association of Software and Service Companies. Available online: *http://www.nasscom.org/*.

National Science Board (NSB). 2002. *Science and Engineering Indicators 2002.* NSB-02-1. Arlington, Va.: National Science Foundation.

National Science Board (NSB). 2004. *Science and Engineering Indicators 2004.* Two volumes (vol. 1, NSB 04-1; vol. 2, NSB 04-1A). Arlington, Va.: National Science Foundation.

National Science Foundation. 2003. *Science and Engineering Indicators—2004,* Appendix Table 2-31. Washington, D.C.: National Science Foundation, Division of Science Resources Statistics, Survey of Earned Doctorates, special tabulations.

National Science Foundation. 2004. *Science and Engineering Doctorate Awards: 2003.* Arlington, Va.: NSF 05-300, Division of Science Resources Statistics.

Naughton, Barry, ed. 1997. *The China Circle: Economics and Electronics in the PRC, Taiwan, and Hong Kong.* Washington, D.C.: Brookings Institution Press.

———. 2003. "The Information Technology Industry and Economic Interactions Between China and Taiwan." Revision of a paper presented to conference on New Information Technologies and the Reshaping of Power Relations, CERI, Paris, December 16–17.

Naughton, Barry, and Yuan Qi. 2000. "Inward Technology Transfer and the Chinese National Innovation System." Presented to the U.S.-China Conference on Technical Innovation, Beijing, April 24–25.

Naughton, Barry, and Adam Segal. 2001. "Technology Development in the New Millennium: China in Search of a New Model." In W. Keller and R. Samuels, eds., *Crisis and Innovation: Asian Technology after the Millennium.* New York: Cambridge University Press.

Neelakantan, Shailaja. 2005. "Higher Education Proves No Match for India's Booming Economy." *Chronicle of Higher Education,* 51 (39), June 3.

OECD. 2004. "Potential Offshoring of ICT-Intensive Using Occupations." DSTI/ICCP/IE (19), Paris.

Olson, Gary M., and Judith S. Olson. 2000. "Distance Matters." *Human-Computer Interaction* 15 (2/3): 139–178.

O'Riain, Sean. 2002. "Remaking the Developmental State: The Irish Software Industry in the Global Economy." Ph.D. dissertation, Department of Sociology, University of California, Berkeley.

"Outsourcing: Make Way for China." 2003. *Business Week Online,* August 4. Available online: *http://www.businessweek.com/magazine/content/.* Access date: September 6, 2004.

"Overseas Chinese Head Home." 2001. *China Daily,* January 23. Available online: *http://www.china.org.* Access date: August 21, 2004.

Paltiel, Ari. 2001. "Mass Migration of Highly Skilled Workers: Israel in the 1990s." OECD Seminar on International Mobility of Highly Skilled Workers: From Statistical Analysis to the Formulation of Policies. Paris, June 11–12.

Parthasarathy, Balaji. 2000. "Globalization and Agglomeration in Newly Industrializing Countries: The State and the Information Technology Industry in Bangalore, India." Ph.D. dissertation, Department of City and Regional Planning, University of California, Berkeley, spring.

———. 2002. "An Asian Silicon Valley in Bangalore? Evidence from the Changing Organization of Production in the Indian Computer Software Industry." Unpublished paper, Indian Institute of Technology, Bangalore.

Parthasarathy, Balaji, and Yuko Aoyama. Forthcoming. "From Software Services to R&D Services: Local Entrepreneurship in the Software Industry in Bangalore, India." *Environment and Planning A.*

Paul, Vivek. 2005. "India Empowered to Me Is . . ." *The Indian Express,* September 4, 2005.

Peattie, Lisa. 1987. *Planning: Rethinking Ciudad Guayana.* Ann Arbor: University of Michigan Press.

Piore, Michael J., and Charles F. Sabel. 1984. *The Second Industrial Divide: Possibilities for Prosperity.* New York: Basic Books.

Porter, Michael E. 1998. "Clusters and the New Economics of Competition." *Harvard Business Review* 76 (6, Nov.-Dec.): 77–90.

Portes, Alejandro. 1996. "Global Villagers: The Rise of Transnational Communities." *The American Prospect* 7 (25, March 1–April 1): 74–77. Available online: *http://www.prospect.org/print/.* Access date: September 23, 2004.

———. 2001. "The Debates and Significance of Immigrant Transnationalism." *Global Networks,* 1: 181–193.

———. 2003. "Conclusion: Theoretical Convergencies and Empirical Evidence in the Study of Immigrant Transnationalism." *International Migration Review* 37(3): 874–892.

PricewaterhouseCoopers. 2004. "China's Impact on the Semiconductor Industry." Global Technology Centre, December 2004. Available online: *www.pwcglobal.com.*

Quan, Xiaohong. 2005. "Multinational Research and Development Labs in China: Local and Global Innovation." Ph.D. dissertation, Department of City and Regional Planning, University of California, Berkeley.

Rapaport, Hillel. 2002. "Who Is Afraid of the Brain Drain? Human Capital Flight and Growth in Developing Countries." Stanford Institute for Economic Policy Research, Policy Brief, April. Available online: *http://siepr.stanford.edu/papers/briefs/.* Access date: September 7, 2004.

Rasiah, Rajah. 2001. "Politics, Institutions, and Flexibility: Microelectronics Transnationals and Machine Tool Linkages in Malaysia." In F. C. Deyo, R. F. Doner, and E. Hershberg, eds., *Economic Governance and the Challenge of Flexibility in East Asia.* Lanham, Md.: Rowman and Littlefield.

———. 2002. "Systemic Coordination and Human Capital Development: Knowledge Flows in Malaysia's MNC-Driven Electronics Clusters." United Nations University, Institute for New Technologies, Discussion Paper No. 2002-7.

Reynolds, Paul D., et al. 2001. *Global Entrepreneurship Monitor, 2000 Executive Report.* Babson Park, Mass.: Babson College.

Roberts, Bill. 2004. "Launching Offshore." *Electronic Business,* April 1. Available online: *http://www.reed-electronics.com/.* Access date: September 7, 2004.

Rock, E. B. 2001. "Greenhorns, Yankees, and Cosmopolitans: Venture Capital, IPOs, Foreign Firms, and U.S. Markets." *Theoretical Inquiries in Law* 2 (2, July): 711–744.

Romer, Paul M. 1986. "Increasing Returns and Long-Run Growth." *Journal of Political Economy* 94 (October 5): 1002–1037.

Roy, Subir. 2004. "Not Quite the World's Silicon Valley." *ZDNet India News,* January 2, 2004. Available online: *www.zdnetindia.com.* Access date: August 19, 2004.

Sabel, Charles, and Jonathan Zeitlin. 1985. "Historical Alternatives to Mass Production." *Past and Present* 108: 133–176.

———. 2004. "Neither Modularity or Relational Contracting: Inter-Firm Collaboration in the New Economy." *Enterprise and Society* 5 (3).

Santa Clara County Office of Human Relations. 2000. "Building Borders in Silicon Valley: Summit on Immigrant Needs and Contributions." San Jose, Calif.: Santa Clara County Office of Human Relations.

Saxenian, AnnaLee. 1999. *Silicon Valley's New Immigrant Entrepreneurs.* San Francisco: Public Policy Institute of California.

———. 2002. *Local and Global Networks of Immigrant Professionals in Silicon Valley.* San Francisco: Public Policy Institute of California.

———. 2003. "Government and Guanxi: The Chinese Software Industry in Transition." Discussion Paper, Center for New and Emerging Markets, London Business School.

Saxenian, AnnaLee, and Chuen-Yueh Li. 2003. "Bay-to-Bay Strategic Alliances: Network Linkages Between Taiwan and U.S. Venture Capital Industries." *International Journal of Technology Management* 25 (1 and 2), entire issues.

Saxenian, AnnaLee, and Xiaohong Quan. 2005. "China." In Simon Commander, ed., *The Software Industry in Emerging Markets.* Northhampton, Mass.: Edward Elgar.

Schram, Art. 2001. "Venture Guru." *siliconindia,* September, 66–68. Available online: *http://www.siliconindia.com/Magazine/.* Access date: September 7, 2004.

Science Park Administration. *Science-Based Industrial Park, 1998.* Hsinchu, Taiwan, 1999. Available online: *http://www.sipa.gov.tw/en/.*

Scranton, Philip. 1983. *Proprietary Capitalism: The Textile Manufacture at Philadelphia, 1800–1885.* Cambridge: Cambridge University Press.

———. 1997. *Endless Novelty: Specialty Production and American Industrialization, 1865–1925.* Princeton, N.J.: Princeton University Press.

"Searching for Hope: Special Semicon West Show Report." 2002. *Solid State Technology,* September.

Segal, Adam. 2003. *Digital Dragon: High Technology Enterprises in China.* Ithaca, N.Y.: Cornell University Press.

"SEMICON China 2004—Speakers at SEMI Market Briefing Highlight Opportunities, Challenges." 2004. March 19. Available online: *http://www.smics.com/.* Access date: September 14, 2004.

Sen, Pronab. 1994. "Software Exports from India: A Systemic Analysis." *Electronics Information and Planning* 22, 2: 55–63.

Sen, Sumantra, and Francine Frankel, eds. 2005. "India's Strategy of IT-Led Growth: Challenges of Asymmetric Dependence." Center for the Advanced Study of India, Philadelphia.

"Shanghai Presents 156 Projects to Overseas Talents." 2004. Silicon Valley Chinese Engineers Association, April 12. Available online: *http://www.scea.org/.* Access date: August 21, 2004.

Shankar, Pradeep, and Karthik Sundaram. 2003. "The Cross-Border Nadkarni." *siliconindia,* February 2. Available online: *http://www.siliconindia.com/.* Access date: September 23, 2004.

Shih, Stan. 1996. *Me-Too Is Not My Style: Challenge Difficulties, Break Through Bottlenecks, Create Values.* Taipei, Taiwan: The Acer Foundation.

———. 2000. Presentation at Monte Jade Science and Technology Association (West) Tenth Annual Conference, Santa Clara, California, March 18.

———. 2002. *Growing Global: A Corporate Vision Masterclass.* New York: Wiley.

Simons, Lewis M., and Michael Zielenziger. 1994. "Bringing US High Tech to Land of Low Pay." *San Jose Mercury News,* June 28, p. 13A.

Skanderup, Jane. 2004. "Taiwan's Cross-Strait Strategy and the WTO." Pacific Forum CSIS. Available online: *www.csis.org/pacfor/issues/.* Access date: September 20, 2005.

Solomon, Jay, and Kathryn Kranhold. 2005. "In India's Outsourcing Boom, GE Played a Starring Role." *Wall Street Journal,* March 23, 2005.

Sowell, Thomas. 1996. *Migrations and Cultures: A World View.* New York: Basic Books.

Srinivasan, S. 2003. "Bangalore's Infrastructure Can't Keep Up with the Tech Boom." *InformationWeek,* September 2. Available online: *http://www.informationweek.com/.* Access date: September 7, 2004.

Srinivasan, T. N. 2005. "Information Technology Enabled Services and India's Growth Prospects." Draft for Brookings Institution, May 5.

Stanford Alumni Conference. 2000. Tapei.

Storper, Michael. 1997. *The Regional World: Territorial Development in a Global Economy.* New York: Guilford Press.

Stremlau, John. 1996. "Bangalore: India's Silicon Valley." *Monthly Labor Review* (November 1996), 119(2): 50–52.

Stub, Zev. 2003. "An Optics Maker with Laser-Sharp Vision." *Jerusalem Post,* October 29, p. 16. *http://www.jpost.com/.* Access date: September 7, 2004.

Studwell, Joe. 2002. *The China Dream: The Elusive Quest for the Last Great Untapped Market on Earth.* London: Profile.

Sturgeon, T. J. 2002. "Modular Production Networks: A New American Model of Industrial Organization." *Industrial and Corporate Change* 11(3): 451–196.

Sundaram, Karthik. 2003. "Dham: Beating a 'New Path' in Chip Design out of India." *siliconindia,* February. Available online: *http://www.siliconindia.com/.* Access date: September 7, 2004.

Taipei. 2000. December. MOEA, Monthly Bulletin of Statistics. *National Economic Trends.*

"Taiwan Redefines OEM." 1995. *Nikkei Electronics Asia,* June, p. 73.

"Taiwan's Electronic Devices Production Expands as Moves to China Progress: Fuji-Kezai Survey." 2004. *NE Asia Online,* March 24. Available online: *http://neasia.nikkeibp.com/.* Access date: August 21, 2004.

Tanzer, Andrew. 1998. "Seizing an Opportunity." *Forbes* 161 (11, June 1): 126.

Teubal, M., G. Avnimelech, and A. Gayego. 2000. "Globalization and Firm Dynamics in the Israeli Software Industry." Unpublished paper, August.

"The Flexible Tiger." 1998. *The Economist,* January 3, p. 73.

"The Kohli Touch." 2000. *Computers Today,* April 16. Available online: *http://www.india-today.com/.* Access date: September 7, 2004.

"The Little Start-Up That Could: A Conversation with Raman Roy, Father of Indian BPO." 2004. Knowledge at Wharton—Operations Management. Available online: *http://knowledge.wharton.upenn.edu/articles.* Access date: September 7, 2004.

Thomson Venture Economics. 2004. "Asian Private Equity Holding Steadfast." News Release, Hong Kong, March 9.

Thurm, Scott. 1998. "High-Tech Firms Big and Small Go Global." *San Jose Mercury News,* May 9.

Todaro, Michael P. 1985. *Economic Development in the Third World.* New York: Longman.

Trajtenberg, Manuel. 2000. "R&D Policy in Israel: An Overview and Reassessment." NBER Working Paper No. w7930, National Bureau of Economic Research, Cambridge, Mass. October 2000. Available online: *http://papers.nber.org/papers/.* Access date: September 14, 2004.

———. 2001. "Innovation in Israel 1968–1997: A Comparative Analysis Using Patent Data." *Research Policy* 30 (3): 363–389.

Tzeng, David. 2001. "Sunrex Chairman Discusses Component Manufacturing in China." Digitimes.com, April 30. Available online: *http:/www.digitimes.com/.* Access date: May 15, 2001.

———. 2003. "Taiwan to Benefit from Japanese Notebook Vendors'

Falling Market Share." Digitimes.com, September 15. *http://www.digitimes.com/*. Access date: September 23, 2004.

Tzeng, Gwo-Hshiung, and Meng-Yu Lee. 2001. "Intellectual Capital in the Information Industry." In Chun-Yen Chang and Po-Lung Yu, eds., *Made in Taiwan: Booming in the Information Era.* Hackensack, N.J.: World Scientific.

United Nations Development Programme. 2003. Human Development Indicators, *Human Development Report.* New York: Oxford University Press.

U.S. Bureau of Economic Analysis, U.S. Department of Commerce, Survey of U.S. Direct Investment Abroad, annual series. Available online: *http://www.bea.gov/*. Access date: August 20, 2004.

U.S. Census Bureau. 2003. Census 2000, *Public Use Microdata Samples (PUMS).*

U.S. Embassy, Beijing. 1997. "Bringing the Students Home: Why They Stay, Why They Return." February. Available online: *http://www.usembassy-china.org*. Access date: August 21, 2004.

Valigra, Lori. 1998. "UTStarCom Leverages Its Presence in China." (View from Stateside), News from Asia Pacific. December. Available online: *http://www.nikkeibp.com/*. Access date: September 7, 2004.

Varma, Yograj. 2004. "Overview: Riding the Growth Wave." *DQ Top 20, Dataquest India.* Available online: *www.dqindia.com/*. Access date: May 2, 2005.

Vijaybaskar, M., S. Rothboeck, and V. Gayathri. 2000. "Labour in the New Economy: Case of the Indian Software Industry." *Indian Journal of Labour Economics* 44 (1): 39–54.

Wade, Robert. 1990. *Governing the Market: Economic Theory and the Role of Government in East Asian Industrialization.* Princeton, N.J.: Princeton University Press.

Wagle, Dileep M. 2000. "China's Emerging Private Enterprises: Prospects for the New Century." Washington, D. C.: International Finance Corp.

Waldinger, Roger, and Mehdi Bozorgmehr, eds. 1996. *Ethnic Los Angeles.* New York: Russell Sage Foundation.

Waldinger, Roger, et al. 1990. *Ethnic Entrepreneurs: Immigrant Business in Industrial Societies.* Newbury Park, Calif.: Sage.

Walsh, Kathleen. 2003. "Foreign High-Tech R&D in China." The Henry L. Stimson Center, Washington, D.C. Available online: *www.stimson.org.* Access date: September 29, 2005.

Wang, C. 1999. "The Information Technology Industry @ Taiwan." *Financial Information.* Taipei, Taiwan.

Wang, Jici, and Jixian Wang. 1998. "An Analysis of New-Tech Agglomeration in Beijing: A New Industrial District in the Making?" *Environment and Planning A* 30: 681–701.

Wang, Jici, and Mark Wang. 2002. "High and New Technology Industrial Development Zones." In M. Webber, M. Wang, and Zhu Ying, eds., *China's Transition to a Global Economy.* New York: Palgrave MacMillan.

Wank, David. 1999. *Commodifying Communism: Business, Trust, and Politics in a Chinese City.* Cambridge: Cambridge University Press.

White, Steve, Jian Gao, and Wei Zhang. 2002. "China's Venture Capital Industry: Institutional Trajectories and Systems Structure." Paper prepared for INTECH, International Conference on Financial Systems, Corporate Investment in Innovation and Venture Capital, Brussels, November 7–8.

Wilson, Dominic, and Roopa Purushonthaman. 2003. "Dreaming with BRICs: The Path to 2050." Goldman Sachs Global Economics Paper No. 99, October 1.

Wilson, Ron. 2004. "Gold Rush Days for Global Design?" *Electronic Engineering Times,* March 26. Available online: *http://www.eet.com/.* Access date: September 7, 2004.

Woetzel, Jonathan R. 2004. "A Guide to Doing Business in China." *McKinsey Quarterly.* Available online: *http://www.mckinseyquarterly.com/.* Access date: September 19, 2004.

Woetzel, Jonathan R., and Andrew Chun Chen. 2002. "Chinese Chips." *McKinsey Quarterly* 2. Available online: *http://www.mckinseyquarterly.com/article.* Access date: September 19, 2004.

Wu, Perry. 2005a. "The Myth of Uncertainty with Chinese Portals." February 9. *www.chinatechnews.com.* Access date: February 9, 2005.

———. 2005b. "Henry Blodget Does China." February 17. *www.chinatechnews.com.* Access date: February 17, 2005.

Wu, Ray. 2004. "Making an Impact." *Nature,* Vol. 428, March 11, 2004, p. 206.

Xia, Yinqi. 2004. "China's Science Parks." Talk by Deputy Director, Administrative Committee of Zhongguancun Science Park, Bridging the Digital Divide, University of California, Berkeley, April 2.

Xiao, Wei. 2002. "The New Economy and Venture Capital in China." *Perspectives* 3(6), September 30. Available online: *http://www.oycf.org/.* Access date: September 7, 2004.

Ye, Guo Bai. 2003. "Shanghai Silicon Intellectual Properties Transaction Center Established in China." Xinhuanet.com, October 3. Available online: *http://news.xinhuanet.com/.* Access date: October 10, 2003. (Translated from Chinese.)

Yoshida, Junko, and George Leopold. 2001a. "China's Top Grads Invigorate the Valley." *EE Times,* August 7. Available online: *http://www.eetimes.com/.* Access date: September 7, 2004.

———. 2001b. "Valley Entrepreneurs Look Homeward to China." *EE Times,* August 15. Available online: *http://www.eet.com/.* Access date: September 7, 2004.

Young, Allyn. 1928. "Increasing Returns and Economic Progress." *Economic Journal* 38: 527–542.

Zeitlin, Jonathan, and Gary Herrigel, eds. 2000. *Americanization and Its Limits: Reworking US Technology and Management in Post-War Europe and Japan.* New York: Oxford University Press.

Zhang, Junfu. 2003. "High-Tech Start-Ups and Industry Dynamics in Silicon Valley." Public Policy Institute of California. Available online: *http://www.ppic.org/content/pubs/.* Access date: September 7, 2004.

Zoellner, Tom. 2000. "Doing Business in Dual Worlds." *San Francisco Chronicle,* January 24, pp. B1–B6.

Zuboff, Shoshanna. 1988. *In the Age of the Smart Machine: The Future of Work and Power.* New York: Basic Books.

Zweig, David. 2004. "Redefining the Brain Drain: China's 'Diaspora Option.'" Working Paper No. 1, Center on China's Transnational Relations, Hong Kong University of Science and Technology.

ABBREVIATIONS

AAMA	Asia America MultiTechnology Association (formerly Asian American Manufacturers Association)
ADSL	asymmetric digital subscriber line
AmBAR	American Business Association of Russian Expatriates
ArmenTech	Silicon Armenia
ASE	Advanced Semiconductor Engineering
ASICs	application-specific integrated circuits
ATV	Acer Technology Venture, America
"B=S+1"	Beijing = Shanghai plus one
BPO	business process outsourcing
CASPA	Chinese American Semiconductor Professionals Association
ChinaSoft	China National Computer Software & Technology Service Corporation
CIE	Chinese Institute of Engineers
CINA	Chinese Information and Networking Association
CMM	Carnegie-Mellon SEI capability maturity model certification
CMOS	complementary metal-oxide-silicon
CSPA	Chinese Software Professionals Association

DRAM	dynamic random access memory
E-Club	Entrepreneur Club
EDA	electronic design automation
ERSO	Electronics Research and Services Organization
GE	General Electric
GSMC	Grace Semiconductor Manufacturing Corporation
Hon Hai	Hon Hai Precision Industry
HP	Hewlett-Packard
IC	integrated circuit
ICSI	Integrated Circuit Solution Inc.
IDMs	integrated device manufacturers
III	Institute for Information Industry
IP	intellectual property
IPO	Initial Public Offering
ISO	International Organization for Standardization
ISSI	Integrated Silicon Solution Inc.
IT	information technology
ITRI	Industrial Technology and Research Institute
kbps	kilobits per second
LAN	local area network
MII	Ministry of Information Industry
NACSA	North American Chinese Semiconductor Professionals Association
NASSCOM	National Association of Software and Service Companies
Nortel	Northern Telecom
NPL	nonperforming loans
NRI	non-resident Indian
ODM	original design manufacturing

OEM	original equipment manufacturing
PAS	personal access system
PC	personal computer
PDA	personal digital assistant
R&D	research and development
S&E	science and engineering
SAS	Silicon Automation Systems
SCEA	Silicon Valley Chinese Engineers Association
SEMI	Semiconductor Equipment and Materials Institute
SiGe	silicon germanium
SIPA	Silicon Valley Indian Professionals Association
SMIC	Semiconductor Manufacturing International Corporation
SOEs	state-owned enterprises
STAG	Science and Technology Advisory Group
STPs	Software Technology Parks
Suntek	Sun Technology
SV	Silicon Valley
TeNeT	Telecommunications & Computer Networks Group
TI	Texas Instruments
TiE	The Indus Entrepreneur
TSMC	Taiwan Semiconductor Manufacturing Corporation
VAT	Value-Added Tax
VC	venture capital
VLSI	very large scale integration
VPN	virtual private network
WIIG	Walden International Investment Group
WLAN	wireless local area network
WTO	World Trade Organization

ACKNOWLEDGMENTS

I began the research for this book in 1996 when I needed a manageable project after the birth of my second child. I had interviewed several Chinese and Indian immigrants during my earlier research in Silicon Valley, so my goal was simply to learn how many foreign-born scientists and engineers worked in the region, and whether they became entrepreneurs at the same rate as their native-born counterparts. Special thanks to Michael Teitz, who first encouraged me to pursue this research, and to the Public Policy Institute of California for funding both the initial project, entitled "Silicon Valley's New Immigrant Entrepreneurs," and the follow-on survey, "Local and Global Networks of Immigrant Professionals in Silicon Valley."

The project that I had intended to be local quickly became global as I realized that the most interesting aspect of the story was the connections that these immigrants were forging back to their home countries. Over the past decade I witnessed the birth and growth of many of the immigrant professional and technical associations that promote these cross-regional connections. I owe a debt to the leadership of these associations, who helped me by welcoming me to their meetings, waiving fees so that I could attend conferences, urging their members to participate in a survey, and patiently responding to my questions. I cannot list all of them here, but they are featured in the book; their importance should be self-evident.

Graduate students are the heart of a research university, and I could not have completed this project without them. I have relied heavily on the brains, the hard work, the language skills, and the cultural and institutional know-how of my foreign-born graduate students to complete this research. Huge thanks go to Jinn-Yuh Hsu, Michele Li, Jumbi Edulbehram, Guoqiang Yu, Wendy Li, Peter Hall, Xiaohong Quan, Yasuyuki Motoyama, and Kyoung-Mun

Shin. I also learned a great deal by supervising theses on related topics. I would like especially to thank two undergraduates, Tam Bui and Radhika Kunamneni, as well as a master's student, Gerald Autler, who came close to writing a dissertation.

Many thanks to Tim Bresnahan for sponsoring a year's leave as a Visiting Scholar at the Stanford Institute for Economic Policy Research at a critical time in the research, and to our collaborators on the project on growing high-technology clusters. I learned a great deal from our conversations. My colleagues in the Department of City and Regional Planning at UC Berkeley were supportive and patient with my absence during this period, for which I am most appreciative; and I joined Berkeley's School of Information Management and Systems while I was working on the book, and benefited greatly from exposure to an entirely new intellectual community.

I interviewed hundreds of engineers, managers, investors, policymakers, and entrepreneurs for this research; they were located in Silicon Valley, Taiwan, Bangalore, New Delhi, Shanghai, Shenzhen, and Beijing. While I can't acknowledge them individually here, they deserve special thanks for generously sharing both their time and their insights. Colleagues who provided useful feedback on earlier versions of this work include Erica Schoenberger, Gary Herrigel, Mark Granovetter, Michael Piore, Richard Locke, Richard Swedberg, Victor Nee, Michael Storper, Karen Polenske, Ashok Desai, and Ken Keniston. I have been privileged to work with a wonderful agent, Lisa Adams. I am also grateful to Mike Aronson of Harvard University Press for his continuing support of my work.

Finally, I want to acknowledge my close friends and family. Thanks go to Robin Broad and Jessica Broitman, who have tolerated my anxieties and single-mindedness with grace. Chuck Sabel was a good friend and colleague throughout, and I owe him special thanks for suggesting the title. My deepest gratitude is to my three guys: Marty, Jamie, and Robbie. This book would not have been possible without their patience and their love.

INDEX